Integrins &
Anti-integrins

Integrins & Anti-integrins

Links of the Chain:
Autoimmune Disease ∞
Anti-integrins ∞
Virus Disease

Vanadiya Ermekova

iUniverse, Inc.
New York Lincoln Shanghai

Integrins & Anti-integrins
Links of the Chain:
Autoimmune Disease ∞ Anti-integrins ∞ Virus Disease

iUniverse books may be ordered through booksellers or by contacting:

iUniverse
2021 Pine Lake Road, Suite 100
Lincoln, NE 68512
www.iuniverse.com
1-800-Authors (1-800-288-4677)

ISBN-13: 978-0-595-40145-1 (pbk)
ISBN-13: 978-0-595-84525-5 (ebk)
ISBN-10: 0-595-40145-7 (pbk)
ISBN-10: 0-595-84525-8 (ebk)

Printed in the United States of America

I dedicate this book to the loving memory of my father and mother.

"Adding one thing to another to discover the scheme of things—"
—Ecclesiastes,
Wisdom, 7:27
The Holy Bible, New International Version
The Zondervan Corporation, 1984:501

Content

Preface

The tremendous progress in the biological sciences leads to remarkable advances in the understanding of the molecular mechanisms of a number of diseases and consequently, in the therapeutic approaches to those diseases. The interpretation of disorders in the light of the latest scientific discoveries resulted in a great evolution in medical science. We now have numerous modern resources in our disposal—novel ways of investigating, diagnosing, and treating ailments with drugs that can control many illnesses or, at least, relieve their symptoms. But the vast improvements in today's medicine and the widening opportunities for pharmaceutical treatment are coming with sometimes deadly threats. New therapeutic agents can bring along unexpected side effects. More and more often, it becomes evident that adverse events, and especially late complications are directly associated with treatment.

One of the reasons for unforeseen complications in treatment is that the recent scientific findings on which the development of new drugs is based only consider one side of the possible effects of the agent, but do not extend to the entire scope of molecular mechanisms the drug's actions. It is understandable that continuous progress in the pharmaceutical industry cannot stop and wait for the possible future discovery of unsuspected, hidden, or even inexistent additional mechanisms that would support, broaden, or change our knowledge and clarify every aspect of the drug action. But sometimes seemingly disconnected experimental facts, which have already obtained in different scientific areas, have not been completely assimilated and analyzed. Hence, solid and available scientific information has not been used for the theoretical assessment of the drug benefit/risk ratio. As a result, the multifaceted mechanisms of the drug action, which were not taken into account during the development of the drug, may influence physiological processes and may be manifested in unpredictable dangerous complications.

Among the major achievements of fundamental science was the discovery of the integrin receptor family, or cell adhesion molecules. Since this discovery, around 20 years ago, extensive research in various areas of academic and medical science have shown that integrins are expressed on the surface of multiple cell types and have proved that integrin-mediated cell-to-cell and cell-to-extracellular matrix adhesion is an important step in the regulation of many biological processes, including immune response, homeostasis, and embryogenesis. There was an attempt of a rapid transfer of the scientific findings in the integrin field to practical applications. One of the pioneering integrin-antagonist drugs is the anti-integrin α_4 monoclonal antibody, natalizimab (Tysabri). Natalizumab (Tysabri) was reported to be a very effective agent for multiple sclerosis and Crohn's disease, the beneficial effect of which is mainly caused by the binding of the anti-integrin α_4 monoclonal antibody to integrins on the surface of activated immune T cells, and resulting in the blockage of T cells trafficking into the sites of inflammation. At the same time, we have to be aware that the anti-integrin α_4 monoclonal antibody's antigen recognition is not limited only to the α_4 integrins expressed on the surface of activated immune T cells, but this antibody can also recognize all α_4 integrins located on the surface of various cell lineages. Besides controlling the adhesion and migration of immune T cells, these integrins are involved in numerous other vital cell processes that can be disregulated by the anti-integrin monoclonal antibody. This necessitates the most careful assessment of the benefit/risk ratio during the any treatment of autoimmune conditions with natalizumab (Tysabri).

This book is written for biologists and medical doctors, particularly for those who wish to follow the progress outside to their own specialized fields. My goal is to integrate the findings pertaining to several areas in fundamental and medical science and to show that there are the links in the same chain: autoimmune disease multiple sclerosis; topographical and functional diversity of adhesion molecules; immunosuppression; viral disease, i.e., progressive multifocal leukoencephalopathy; and molecular biology of JC virus.

The structure of the book

Each chapter consists of an introductory part, which sets forth basic principles and information about the topic. The body of the text is subdivided into several sections. Each section culminates with "key points" emphasizing the factual material. At the end of certain sections, there are the author's opinion and conclusions aiming in part to reinforce the most important information.

A brief list of references at the end of each section is arranged in chronological order. Naturally, it is impossible to refer to all available publications in a book which reviews several areas of science. I have mostly resorted to the earlier and latest original papers in which important discoveries were first reported and to the comprehensive reviews, where readers who wish to go beyond the text can find secondary references. But even this list had to be severely restricted, and I extend my apologies to authors whose essential data and publications I have omitted. My choice of references was not dictated by their support for "link of the same chain" concept; on the contrary, I discussed a number of opinions that I found in the literature. Elsewhere in the book, I have avoided naming individual researchers, but nevertheless, several names have been mentioned for two reasons. First, I was so impressed with some sophisticated scientific experiments and the interpretation of them that I could not help but mention author's name; second, I wanted to identify the investigators of particular clinical trials in order to facilitate the search for the corresponding references.

The majority of terms and abbreviations are explained in the text. A brief glossary of terms unexplained in the text and a list of the abbreviations are given at the end of the book. Definitions of biological terms are provided for medical doctors, and descriptions of medical terms are provided for biologists, both of whom may encounter nomenclatures that require more clarification.

The book contains 14 figures. There are no original photos or electron micrographs but only schematic figures. No graphics can adequately capture the enormous complexity of cell anatomy and functions but can serve only as an approximate explanation of intracellular processes. Therefore, I tried to make the visual illustrations as simple, clear, and understandable as possible.

Although the chapters of this book can be read independently of one another, I hope that the reader will see that they are arranged in a logical sequence which closely follows the concept of "links in the same chain":
Multiple sclerosis ∞ *treatment by integrin antagonists* ∞ *immunosuppression* ∞ *viral disease.*

Chapter 1, "Immunity; Normal and Altered Functions of Immune System," contains an introduction to the basic biology of the immune system and the mechanisms of self-tolerance and autoimmunity. The chapter as a whole describes the clinical course of multiple sclerosis, pathophysiology, and the conventional FDA-approved treatment for the disease.

Chapter 2, "Integrins—Cell Adhesion Molecules," characterizes the structure and functions of integrins, with a special emphasis on the integrin's topographical and functional diversity. This material is augmented by information about the bidirectional signaling of integrins, which leads to their involvement in the regulation of a number of cell processes.

Chapter 3, "Monoclonal Antibody Therapy," in its introductory part consolidates the general principles of monoclonal antibody development and characterizes the currently manufactured monoclonal antibody types. It then focuses on the anti-integrin monoclonal antibody, natalizumab (Tysabri), as a treatment for autoimmune diseases, i.e., multiple sclerosis and Crohn's disease. This part contains the most complete and detailed data analysis of the trials that have been conducted with natalizumab (Tysabri). To urge a more critical evaluation of clinical trial data, the analysis of each trial is provided with key points, as well as the authors' interpretation. Also presented are the milestones of the marketing history of the drug: aggressive promotion before FDA approval; FDA granted accelerated approval based on unpublished data from the first year of two Phase III two-year studies (November 23, 2004); withdrawal from the market following reports of the two confirmed cases of severe adverse events—progressive multifocal leukoencephalopathy—directly associated with the treatment (February 28, 2005); and submission of a supplemental Biologic License Application to the FDA, with a request for

Priority Review, to resume marketing of natalizumab (Tysabri) (September 26, 2005).

Chapter 4, "Progressive Multifocal Leukoencephalopathy," describes the clinical course, etiology, and pathogenesis of the progressive multifocal leukoencephalopathy and focuses on immunosuppression that is a precondition for the development of the progressive multifocal leukoencephalopathy and a prognostic factor for the outcome of the disease.

Chapter 5, "Viruses," in its introductory part provides general information about viruses, and then goes on to elucidate the behavior of JC virus, etiological agent of progressive multifocal leukoencephalopathy, with particular emphasis on the three different modes of JC virus life cycle.

Chapter 6, "Summary and Reflections," synthesizes the data and stresses the functional diversity of integrins. It also outlines my thoughts and interpretations of the benefits and limitations of integrin-antagonist treatment.

Acknowledgement

This book could have not been written without the support of many people. I am grateful to everyone who helped me with this undertaking.

Most of all I want to thank my daughter, Kira Sheinerman, and my son-in-law, Felix Sheinerman, both Ph.D. in molecular biology, for their enthusiastic encouragement, scientific discussion and criticisms, moral and financial support during my work on the book.

In writing this book I benefited from the great privilege of having many close friends around the world who are talented scientists and experienced medical doctors. They never failed to share with me the profound depth of their knowledge, to provide advice and assistance on the subjects in which they are experts. They don't need to be mentioned by name: they know who they are, and thank all of them very much.

I wish to acknowledge the many brilliant researchers to whom I have referred, whose publications, filled with innovative ideas and sophisticated experiments, engendered in me a strong scientific curiosity about integrins and whose experimental data formed the bedrock of this book.

I would like to thank my former colleagues from the Worldwide Healthcare Communications company. Special thanks to the president of the company, Sid Auerbach, and CEO, Steven Schlackman, who established such a creative environment in the company, and to the experts in computer art design and business management, Kurt Ortell, William Beckwith, Christine Mauro, and Robert Wainwright. They were the first people to whom I presented my initial thoughts during one of our regular company meetings and were the first who encouraged me to emphasize the topographical and functional diversity of integrins and the risks connected with the use of anti-integrin agents. About two years have elapsed since that meeting, but I still remember and very thankful to all of my former colleagues for fostering a challenging atmosphere during discussion.

I thank all my friends, who are biologists and specialists in other areas, and who, perhaps overcoming their boredom, still never failed to listen patiently to the long rounds of drafts and tolerated my false promises that each one of them was the final one.

From the first to the final draft, I had several editors. My sincere gratitude goes to all of them for their highly professional work, and especially to Tanya Preiser for her editorial help, acute observations, and valuable advice. Any errors are completely my fault, because not unlike other writers, I had author's ambition to keep my own style, which sometimes contradicted the editors' recommendations.

I thank again my son-in-law, Felix Sheinerman, who prepared all the computer-designed illustrations. And my very special thanks go to the painter Pavel Tayber for his creative work on the design of the book cover. The idea behind it was that even the most sincere efforts and the most time-consuming labors may still lead only to results that are ambiguous. And Pavel Tayber transformed this idea into a metaphorical (double-faced) oil painting performed with his free and confident artistic skill.

I thank my dear friend Diane De Marchena who was diagnosed with multiple sclerosis, is now fighting the disease, has a productive life, and looks beautiful. I'd like to thank Diane for telling me the story of her disease and endowing me with a patient's insight into multiple sclerosis, which is very important for us—the scientists, drug developers, and other people who plan, conduct, or review clinical trials.

If this book is useful in any way for its targeted audience, I want to share my satisfaction I would feel with all these people.

Chapter 1

Immunity
Normal and Altered Functions of
Immune System

Major Components of Immune System

Through the evolution, the human body has developed an extremely powerful, sophisticated immune system against foreign agents such as bacteria, viruses, xenograft and allograft tissues, cancer cells, etc. Major components of this multi-branch system include both innate and adaptive immunity: the complement system, macrophages, natural killer (NK) cells, B and T lymphocytes, and cytokines.

The complement system is an integral constituent of the innate immunity and one of the most ancient systems of the immunological defense, which eliminates foreign substances from the body. Macrophages have ability to phagocyte microbes, necrotic tissues, and dead cells. NK cells, which previously have never encountered the antigenic targets, such as virus-infected cells, tumor cells, and xenograft or allograft cells, nevertheless, provide immunosurveillance against them. Lymphocytes—B and T cells—are the primary cells of the adaptive immune system, which ensure one the most specific and comprehensive defense mechanisms in the immune system. B cells defend against the extracellular pathogens by producing antibodies (Abs). T cells play a central role in orchestrating the immune response. T cell-mediated immunity includes the participation of several T cell types: cytotoxic T lymphocytes with accessory receptor CD8+ (CTLs), or T-killers, that destroy the foreign agents by the direct attack and eliminate the intracellular pathogens (most viruses and some bacteria); helper T lymphocytes with accessory receptor CD4+ that increase an activation of B cells, CTLs, and macrophages; suppressor T lymphocytes that regulate the activities of other cells keeping them from excessive immune reactions. Dendritic cells play a key role in initiating the immune response by presenting the foreign antigens to T cells. Cytokines have the complex antiviral, antineoplastic, and immuno-regulatory functions. Interferons (IFNs) belong to the cytokine family and are produced in response to the viral infections. For efficient immunity, all branches of the immune system—humoral immunity, cell immunity, and cytokines—have to cooperate.

Causes of Immunosuppression

Causes of immunosuppression may be considered according to the categories schematically outlined below.

The first category includes the certain diseases: AIDS, chronic leukemia and lymphomas, carcinomatoid diseases, and chronic inflammatory diseases (tuberculosis and non-tropical sprue). Some of these diseases, such as AIDS, cause immunosuppression. Other diseases may be developed because of pre-existing low levels of natural immunity, for instance, it is well known that chronic infectious diseases occur mainly due to the decreased immune resistance; and the very existence of malignancy in a person is usually a testimony to the failure of the immune system in dealing effectively with the transformed cells. Moreover, infections are a common cause of morbidity and even death in the patients with a wide range of neoplasms.

The second category consists of the cytotoxic therapy including anti-cancer chemotherapy and radiotherapy. Although anti-cancer therapies are designed to kill tumor cells, the cytotoxic chemotherapeutic agents and radiation almost invariably affect the functioning of the bone marrow cells. Lymphocyte progenitors are very sensitive to the cytostatic therapy and exposure to radiation, sometimes even more so than the malignant cells. Anti-cancer therapy non-specifically destroys large populations of lymphocytes resulting in the secondary immunosuppression.

The third category is comprised of the therapy specifically directed against the functional activity of the immune system cells. This category may be subdivided for two groups. In the first group, the target goal of therapy is immunosuppresion. For instance, immunosuppression is specifically indicated for organ transplant recipients to decrease the frequency and severity of organ allograft rejection. This therapy, as currently available, generally suppresses all immune responses including those to bacteria, fungi, viruses, and even to malignant cells. In less than a month after transplantation, the consequences of the administration of the inhibitors of immune cell functioning become apparent—patients with immunosuppression succumb to opportunistic infections. In the second group, immunosuppression is a logical but undesirable result of the treatment. Two examples clearly demonstrated the appearance of infectious complications after the anti-cancer and anti-inflammatory treatments that have led to the secondary immunosuppression: a). the combination of the anti-cancer monoclonal antibody directed against the CD20+ antigen on B-cells with CD34-selected transplantation; and b). anti-inflammatory immunosuppressive therapy for the treatment of relapsing forms of multiple sclerosis utilizing the monoclonal antibody against α_4 integrin both resulted in the impairment the immune functions and in the increase rates of bacterial and serious viral infections.

References

Goldberg et al, 2002; Boye et al, 2003; Biogen Idec, Elan Pharmaceutic Press release, 2005; Carpenter et al, 2005; Kleinschmidt-DeMasters and Tyler, 2005; Langer-Gould et al, 2005; Van Assche et al, 2005.

Key Points:

1. The interactions among the components of the innate and adaptive immune systems constitute the fundamental defense against viral infections.

2. Resistance to infectious diseases requires a certain level of the immune system functionality.

3. Immunosuppression leaves a patient vulnerable to infectious diseases.

4. Immunosuppressive drugs must be used judiciously with strict attention to the hazards of promoting severe and fatal infections and malignancy. Generalized pharmacological immunosuppression must be resorted to only as a "rescue" therapy.

Mechanisms of Self-tolerance

The immune system is able to recognize the foreign substances as "non-self" and to initiate the immunological responses against them while ignoring one's own body components, which are recognized as "self." The immune system ability to distinguish self from non-self and refrain from attacking its own body tissues is called "self-tolerance."

Although the recent discoveries in cell and humoral immunology and molecular biology made a great impact on the elucidation of self-tolerance mechanisms, the main postulates of this phenomenon associated with names of famous immunologists such as R.D. Owen, F.M. Bernet, J. Lederberg, P.B. Medawar, and others who proposed mechanisms of self-tolerance in the middle of the 20th century are still upheld today. In 1945, R.D. Owen first suggested that self-tolerance occurs because of the immune system exposure to the self-antigens during fetal life. Several years later, F.M. Burnet came up with the clonal selection theory. He claimed that each lymphocyte clone is specific for one antigen only; and if a lymphocyte meets this antigen during early formation, the lymphocyte would become a "forbidden clone" and would be deleted. In addition, J. Lederberg stressed the importance of lymphocyte's ontogeny stage. He stated that the stage of lymphocyte maturation at which it encounters the self-antigen is more crucial than the stage of embryogenesis. These fundamental principles, which were further strengthened by the experiments at the molecular level and new immunological findings, are summarized below.

Clonal Deletion

In order to protect the body's own tissues from the immune reactions, the lymphocytes which recognize and target self-antigens must be eliminated. The process of self-tolerance formation is subdivided into an establishing of central tolerance and peripheral tolerance. Central tolerance refers to the mechanisms of tolerance that act during an early lymphocyte development in the primary immune organs—the bone marrow (BM) and the thymus. The immune cells that recognize self-antigens are eliminated before they become fully immuno-competent lymphocytes. The peripheral tolerance refers to the mechanisms of tolerance that act after the mature lymphocytes have left the primary lymphoid organs. The process of peripheral tolerance may include the deletion of immune cells, functional inactivation via anergy (lack of an immune cell activity without clonal deletion), or immune suppression regulated by a variety of mediators.

The central tolerance involves the deletion of high-affinity self-reactive lymphocytes, whereas the peripheral tolerance mechanisms act mainly at the level of low-affinity self-reactive immune cells. The deletion of self-reactive cells at an early stage of differen-tiation was termed "clonal deletion."

The majority of B and T cells, which recognize self-proteins, are eliminated through the genetically programmed cell death, or apoptosis.

Clonal Ignorance
The deletion of lymphocytes specific for self-antigens is the major mechanism of self-tolerance, but it is not the only one. Both T and B cells specific for some self-antigens can be identified in healthy people, and the mechanisms must exist to prevent such cells from becoming activated and damaging for self-tissue. One well-recognized mechanism is called "clonal ignorance." The

clonal ignorance occurs when certain self-proteins are not exposed to the immune system during the embryonic development because they are hidden from the naïve primary immune organ lymphocytes in privileged sites. The privileged sites include eyes, brain, and testis. If during an adult life there was no contact between the mature lymphocytes and these hidden self-antigens, no productive immune response will occur. Having had no exposure to the hidden self-antigens, the immune cells take on the appearance of self-tolerant cells. However, if exposure of the immune cells to the previously hidden self-antigens occurs (for example, due to trauma of the privileged sites), the immune response is dramatic.

It is well understood that the phenomenon of self-tolerance is not a passive silencing of the immune system function, but it is an active process regulated by multiple factors and complex interactions. These include but are not limited to the state of maturity of the lymphocytes and the site of encounters of the immune cells with self-antigens, the host's developmental stage, the sub-classes of lymphocytes (recent findings strongly suggest that $CD4^+$ T cells and T suppressor cells play a crucial role in self-tolerance), and many other factors. Although the mysteries of self-tolerance have been a subject of intense investigation for over 60 years and had a history of great discoveries including Nobel Prize awards, a number of questions still remain unanswered, and the elucidation of the nature of self-tolerance merits further exploration.

References

Miller, 2001; Morshed et al, 2001; Platt and Lakkis, 2001; Goodnow et al, 2005; Kronenberg and Rudensky, 2005.

Mechanisms of Autoimmunity

Unfortunately, the loss of immune self-tolerance is not uncommon. The immune reactions against self-antigens are known as autoimmunity. Autoimmunity results from the malfunctioning of one or more of the basic mechanisms regulating immune self-tolerance. There is a number of different ways in which the breakdown of self-tolerance may occur:

- Any defect in the elimination of forbidden immature lymphocyte clones within the prime organs or an alteration of the peripheral tolerance process may cause autoimmunity.

- Failure to induce apoptosis in the immune cells may lead to autoimmunity. For instance, a mutation in the genes that encode cell apoptosis signals can be an important factor responsible for the loss of self-tolerance and the development of autoimmunity.

- If self-proteins normally hidden from the immune cells during fetal period (such as those in the eyes, brain, and testis) are released into the circulation during an adult life, they may be recognized as non-self by the body immune system. These proteins may elicit an immune response normal for the foreign antigens, but in this case, comprises an autoimmune response. For example, trauma in one eye may induce the release of eye specific proteins from the privileged site, which normally sheltered from the immune cells. The resulting autoimmune response may cause the damage and blindness to both eyes because eye specific proteins in both eyes become marked for attack.

- Under some circumstances, the body's own proteins may undergo conformational changes. Normally acting immune system cannot distinguish them from foreign antigens. The failure in recognition of self-proteins results in an autoimmune response.

- The lymphocytes specific for self-antigens, which are normally inactive, may become activated by "superantigens." Some bacteria and viruses are superantigens capable of eliciting a massive immune response: the release of large amounts of inflammatory cytokines and the stimulation of T and B cells, including self-reactive T and B cells.

- Certain epitopes are shared by both foreign antigens (from bacteria or viruses) and endogenous proteins of the host. Due to this "molecular mimicry," antibodies against shared antigens cross-react with self-proteins and target them for the destruction.

References:

Morshed et al, 2001; Bernet et al, 2003. Goodnow et al, 2005; Kronenberg and Rudensky, 2005; Rioux and Abbas, 2005.

Key Points:

1. The immune self-tolerance is regulated by multiple factors and interactions.
2. Malfunctioning of any of the mechanisms regulating the immune self-tolerance, which may be caused by endogenous or exogenous factors, can lead to autoimmune diseases.

Autoimmunity comprises more than 75 chronic illnesses targeting almost all of the body's organ systems. The following describes in more detail one of the most prevalent and dangerous autoimmune diseases—multiple sclerosis.

Autoimmune Disease—Multiple Sclerosis

Multiple sclerosis (MS) is a chronic autoimmune neurological disease characterized by a cycle of inflammation in the central nervous system (CNS), multiple foci of demyelination, and scarring within the white matter of the CNS.

Symptoms

The clinical manifestations of MS result from a deceleration of the neuronal signal conductivity to the muscle cells. Neurons conduct messages in the form of impulses. All neurons have the same structure: they are composed of the cell body (soma), the metabolic center of the cell; the dendrites, the miniscule extensions receiving incoming information; and the axon, the conductor that transmits outgoing impulses from the neuron body to the target cells.

Each axon is coated with a layer called myelin sheath. Myelin is a lipid-rich membrane, which contains a relatively small amount of proteins. Myelin functions as a living electrical insulator facilitating high speed of the nerve signal propagation along the axons. The length of axons varies from a few microns up to three meters. The neuronal signal propagation rate is directly proportional to the axon diameter. In myelinated axons, the conduction rate in meters per second (m/sec) corresponds approximately to six times the axon's diameter in microns. For example, in myelinated axons, the range of signal propagation rates is between 30 and 120 m/sec. In demyelinated axons of the same diameter, this range is 15 to 20 times slower. Myelin is produced by oligodendrocytes.

MS is characterized by the dissemination of demyelination among various parts of the brain and spinal cord. The type and severity of symptoms depend on the location of the demyelinated lesions and the degree of demyelination. The symptoms include unsteady gait, weakness of limbs, spasticity, rapid involuntary movements of the eyes, optic neuritis (which generally presents itself as diminished visual acuity, dimness, or decreased color perception), double vision, defects in speech pronunciation, and sensory and mental symptoms.

Clinical Patterns

Four clinical patterns are identified by the International Consensus based on the multiplicity of neurological deficits, the sequence of relapses and remissions, the course of disease progression, and the severity of symptoms:

- Relapsing-Remitting MS (RRMS)
RRMS is characterized by relapses followed by complete recovery or by symptomatic episodes of relapses accompanied by a residual deficit upon recovery. Although RRMS is not a severely progressive form of the disease, accumulated irreversible deficits may result in permanent clinical impairment. At initial stages of the disease, RRMS is a typical diagnosis of approximately 85% of all MS patients.

- Secondary progressive MS (SPMS)
SPMS is characterized by a gradual progression of disability with or without intervening relapses. Approximately 50% of the patients with RRMS convert to SPMS within ten to fifteen years since the disease onset.

- Primary progressive MS (PPMS)
PPMS is characterized by a gradual progression of the disability without relapses since the onset of disease, occasional plateaus,

and periods of temporary improvement. PPMS occurs in approximately 10% to 15% of all MS patients.

• Progressive relapsing MS (PRMS)
PRMS is characterized by a gradual progression of the disability with acute relapses since the disease onset. PRMS, the least common form of MS, occurs in approximately 5% of all MS patients.

Diagnosis

Diagnostic criteria of MS include:

• Clinical presentation
• Disease-related magnetic resonance imaging (MRI) changes, which are present in more than 95% of the patients
• Slow or abnormal electrical conductivity, which is detected in 80% to 90% of the patients by evoked potential test
• Two or more oligoclonal bandings (OCBs) in the cerebrospinal fluid (CSF), which determines the levels of intrathecal IgG production, and are detected in 75% to 90% of MS patients

Prognosis

MS affects around 350,000 Americans and 2.5 million individuals worldwide. About 1% of MS patients die every year according to data of the World Health Organization (WHO). Studies with MS patients without treatment (natural course of disease) indicated that all patients with MS develop disability. Even among the patients diagnosed with "benign" forms of MS during the first ten years, half had significant disability twenty years after the time of diagnosis. By fifteen years after diagnosis, less than 20% the patients remained without functional limitations; approximately 50% to 60% the patients required assistance with ambulation; 70% the patients were limited or unable to perform basic daily activities, and 75% the patients were not employed.

References

Lublin and Reingold, 1996; Hawkins and McDonnell, 1999; Satkuman, 2003; Hauser and Goodin, 2005.

Etiology

The etiology of autoimmunity in MS, though unproved, is considered to be influenced by several components: the geographical zone; the exposure to one or more environmental factors including chemical pollutants or infectious pathogens; the genetic vulnerability; and the alterations in one's own immune system with subsequent development of the immune responses against self-proteins in the CNS.

Clusters of MS Distribution

In the course of intense epidemiological studies, it was discovered that MS distribution in the geographical zones varies. There are areas, "hot spots," of MS prevalence. It was noticed that the incidence of MS increases with an increasing distance from the equator (for instance, 250 MS cases per 100,000 were diagnosed in Orkney, Scotland, 2 per 100,000 were identified in Japan, and 15 per 100,000 were reported among the Japanese Americans).

Nevertheless, the epidemiological studies showed that at the same latitude and in the adjacent regions of the same country the incidence rates of MS were quite different. Studies in Finland showed that during 1979–1993 the incidence of MS was 5.1 per 100,000 person-years in the southern district of Uusimaa, 5.2 in the coastal region Vaasa, and 11.6 in Seinajoki, a western district. Vasterbotten County in northern Sweden turned out to be a higher risk area for MS compared to other Swedish counties. In other words, there were the identifiable regional clusters of a high risk incidence of MS. Other reports suggested that the individuals who moved from a territory of high prevalence to one of low prevalence (or vice versa) before the age of 15 adopted the risk of MS in their new environment, whereas if they moved after this age, they retained the risk level of their native land. These findings partially disproved the hypothesis that the disease distribution followed latitude-related gradient only and pointed to the importance of the environmental factors (chemical pollutants and infectious agents) in MS etiology.

Chemical Agents

Since the discovery of MS around the mid-XIX century, there were endless speculations about the role of external factors in the disease. The first reported appearance of the disease occurred in 1826, only a few years after mercury began to be used as amalgam fillings in dentistry. According to the theory existing at that time, a term "oral galvanism" was coined to suggest that mercury in crowns, braces, and fillings strongly affected the patients' CNS, and MS was even called a "dental disease." Recently, there were reports of higher than average rates of MS incidence in Herculanium, in Jefferson County, MO, which was suspected to be linked to the environmental chemical pollution, lead contamination.

Viruses

Although investigators did not consistently isolate a specific bacteria or virus from the tissue of MS patients as a definite etiological agent, substantial amount of evidence suggests that there is an infectious component in the development of MS:

- To date, a statistical correlation was determined between the incidence of MS and infection by certain bacteria and viruses, including viruses of poliomyelitis, measles, herpes simplex, varicella, rubella, Epstein-Barr, and influenza C.

- A number of epidemic reports supported a theory of primary viral infection in the etiology of MS. The most impressive of those was the one about MS epidemic on the Faeroe Islands off the coast of Denmark after the British occupation during Word War II.

- Furthermore, the likelihood of a viral origin in MS is supported by the fact that most demyelinating diseases were conclusively proven to have a viral etiology.

- The idea of a viral etiology in MS is also enhanced by the preponderance of CD8$^+$ lymphocytes in demyelinated areas because it is well known that CD8$^+$ T cells control and clear out viruses at the early phases of intracellular virus multiplication.

- Viral involvement in MS etiology is further substantiated by virus-induced demyelination in animal experimental models of MS.

- Measurable therapeutic effects of anti-viral agent, IFN-β, also pointed to the virus involvement in MS etiology.

As described above, the infectious agents may create autoimmune response acting as superentigens or by the process called molecular mimicry. It was suggested, though not proven, that viruses in MS may function as superantigens increasing the production of inflammatory cytokines and activating the lymphocytes. Another possibility is that molecular mimicry between virus and myelin proteins may trigger the cross-reactivity to myelin components.

The activation of the immune system by superantigens and virus molecular mimicry may lead to autoimmunity and MS.

Genetic Susceptibility

It is generally accepted that MS is an autoimmune disease in genetically predisposed individuals.

Numerous examples from European epidemiological studies showed the potential involvement of genetic mechanisms in MS development: the incidence rates of MS in France are sufficiently lower than those expected based only on its geographical position; MS is less common in Gypsies than in other whites living in the same areas; and exceptionally high familial clustering of the disease is detected among Finns. These data suggest that besides the geographical determinants and environmental agents, the ethnic factors are also likely to be important in MS development.

In addition to data about increased risk for MS with respect to the settlements, migration, and ethnic origin, certain molecular studies implicated genetic mechanisms in the development of MS. The identification of myelin basic protein gene haplotype in MS patients from high-risk area in Western Finland suggested genetic founder effect. Genome-wide linkage analysis performed on 22 Finnish multiplex MS families identified the potential disease predisposing loci on chromosome 17. The study, which examined the progression to the disability in RRMS patients carrying an ancient mutation in the Mediterranean fever gene (MEFV), showed that the carriers of MEFV gene had twice the expanded disability status scale (EDSS) score of the non-carriers at the comparable duration of the disease. The median time to reach an EDSS score of 3 was two years in carriers versus ten years in non-carriers, and the median time to reach an EDSS score of 6 was six years in those harboring a founder mutation compared to twenty three years in those without gene mutation. Family members of MS patients are at increase risk for the disease. Twin studies demonstrated the concordance rate of 25% to 30% in monozygotic twins. At the same time, the fact that not all identical twins will both develop MS suggests that the disease is not under direct genetic control. Genes that cause MS in genetically vulnerable individuals were not identified.

References

Hafler, 1999; Pozzilli et al, 2002; Saarela et al, 2002; Scarisbrick and Rodriquez, 2002; Fazakerley and Walker, 2003; Kenealy et al, 2003; Shinar et al, 2003; Swanborg et al, 2003; Sundström et al, 2004; Tienari et al, 2004; Hauser and Goodin, 2005; Rioux and Abbas, 2005.

Key points:

1. Significant differences in secular trends for the incidence of MS in various geographical areas may be associated with environmental pollutants.

2. Viruses were hypothesized as possible etiological factors for MS.

3. The genetic analyses pointed to the involvement of genetic mechanisms in MS.

4. The most probable scenario is that the development of MS occurs with a combination of external provocative factors in genetically susceptible individuals.

Pathophysiology

Although the initiating agents in the etiology of MS remain unknown, the role of the immune system in the pathogenesis of MS is evident. It is generally accepted that the pathophysiology of MS results from the activated T cells and macrophages. Immune cells cross the blood-brain barrier (BBB), accumulate in the CNS, attack self-antigens, and trigger a broad spectrum of inflammatory events responsible for the brain damage.

Multifactorial Patterns of MS

Variations of MS manifestations suggest the multifactorial rather than solitary pattern of MS pathophysiolgy. This suggestion was recently proven. Several distinct histological and molecular patterns of actively demyelinated lesions were classified on the basis of the immune cell involvement, evidence of autoantibodies deposition, complement activation, myelin protein loss, plaque distribution, and types of oligodendrocyte destruction.

The histological analysis of lesion tissues from 51 biopsy and 32 autopsies specimens performed by Lucchinati and colleagues detected four different patterns of demyelination. All four patterns (I, II, III, and IV) of these lesions had evidence of the inflammatory infiltration by macrophages and T lymphocytes. Two patterns (I and II) showed similarities in the compositions of macrophages and T cells, but pattern II had more prominent deposition of humoral immune components—autoantibodies and active factors of the complement system. Two other patterns (III and IV) definitely pointed to the primary oligodendrocytic dystrophy reminiscent of virus or toxin-induced demyelination. Remarkably, the patterns of demyelination were heterogeneous in different patients but were homogeneous within multiple active lesions of the same patient, which indicated to the different MS etiologies for different patients.

Traditionally, MS is viewed as a disease involving CNS inflammation resulting in axonal demyelination. But now from pathohistological and neuroimaging data it is clear that the damage of the axons and neuron bodies is also a part of MS process. Acute axonal injury correlates with the rate of macrophage and CD8+ T cell infiltration. Demyelination eventually leads to the "shorting" of the axon conductive core and to its destruction. Moreover, it is believed that loss/shrinkage of axons is a major contributor to the brain and spinal cord atrophy, which is considered as a neuroanatomic substrate of MS progression and permanent disability. The MS affected brains show 30% to 35% fewer neurons than control brain tissues. Changes of neuron bodies were detected even in the patients at early MS stages. There were speculations that a neurodegeneration occurs due to the inflammatory injuries as well as from a variety of toxic factors.

Despite the diverse patterns of demyelination, the final events in the pathophysiology of MS are the same, defined as the pathophysiological triad of MS: CNS inflammation with increased numbers of T lymphocytes and macrophages, demyelination, and eventual gliosis (scarring) in the brain.

Immune Components in MS Pathogenesis and Progression

The invasion of activated lymphocytes and macrophages into the brain and subsequent severe inflammation are generally considered as the most important steps in the pathogenesis of MS. At the same time, almost all humoral and cell constituents of the immune system are altered in MS development and progression.

- The breakdown of the BBB takes place at the early stage of MS. It is possible that collapse of the BBB is actually the earliest step preceding all other MS-associated events. Then chemicals, viruses, bacteria, toxins, and formed blood elements are free to permeate the brain. The penetration of the BBB by these normally prohibited invaders may provoke the cascade of autoimmune events.

- The two major myelin membrane proteins, protein 2', 3'-cyclicnucleotide 3'-phosphodiesterase (CNP) and basic myelin protein (BMP), evoke production of the autoantibodies. MS is characterized by intra-BBB synthesis of autoantibodies predominantly against these glycoproteins.

- The autoantibodies bind to the myelin-associated proteins and develop the immune complexes. Depending on the clinical stage of the disease, MS patients exhibit circulating antibody-CNP immune complexes in 83.33% to 100% of all analyzed cases, which is much higher than in the control population.

- Complement system component (C3) binds to the aforementioned antibody-CNP immune complexes providing a plausible mechanism for the opsonization of the myelin membrane. The phagocytosis of antibody-CNP complexes is mediated by macrophages and CNS microglia.

- The number and functional activity of NK cells are significantly reduced in MS patients. Profound deficiency, phenotype defects, and altered functions of NK cells are associated with clinical relapses and disease progression in RRMS patients. The restoration of NK cell functional activity is associated with MS remission. There is a statistically significant reverse correlation between NK functional activity and the total number of active lesions on MRI. It is important to note that the formation of new active and new enlarging lesions is preceded by a reduction in NK functional activity. The experimental results suggest that periods of reduced NK functional activity represent the periods of the susceptibility for the formation of active lesions and clinical attacks.

- The alteration of the cytokine network accompanied by the release of inflammatory cytokines/chemokines occurs in the MS brain. It is presumed that tumor necrosis factor, IFN-γ, and inflammatory interleukins may injure the myelin membrane, which directly contribute to MS pathogenesis.

References

Vranes et al, 1989; Cojocaru et al, 1992; Munscheauer et al, 1995; Kastrukoff et al, 1998; Mun-Bryce and Rosenberg, 1998; Tayeben et all, 1998; Wagstaff and Goa, 1998; Walsh and Murray, 1998; Waunbaunt et al, 1999; Illes et al, 2000; Lucchineti et al, 2000; Goodin, 2001; Takahashi et al, 2001; Chard et al, 2002; Dalton et al, 2002; Genain et al, 2002; Giannelli et al, 2002; Herdon, 2002; Kuhlman et al, 2002; Silber et al, 2002; Bernet et al, 2003; Kastrukoff et al, 2003; Kieseir and Hurtung, 2003; Kraus et al, 2003; Minagar and Alexander, 2003.

Key points:

1. Almost all branches of the immune system are involved in MS pathogenesis.

2. Several components of the immune system are definitively implicated in the pathogenesis of MS. Other immune constituents may play a secondary role maintaining the altered activity of the immune processes in MS.

3. An imbalance in the complex interactions between components of the immune system may be considered as the immunological hallmark of MS.

Treatment Options

Current pharmacological strategies in MS treatment aim to limit demyelination by the reducing inflammation and modifying the immune response.

Since 1993, the US Food and Drug Administration (FDA) approved several disease-modifying therapies, known as immunomodulating agents, for the treatment of MS. These include agents based on interferon-beta (IFN-β), glatiramer acetate (Copaxone), and mitoxantrone (Novantrone). Three IFN-β drugs for patients with MS are approved in the USA: IFN-β 1b (Betaseron) manufactured by Berlex was approved in 1993; IFN-β 1a (Avonex) produced by Biogen was approved in 1995; and IFN-β 1a (Rebif) produced by Serono/Pfizer was approved in 2000. With the introduction of IFN-β agents, the therapy for MS patients was improved, and now they constitute the basic immunomodulatory treatment for patients with RRMS. Among the treated patients with RRMS, 90% received one of the IFN-β drugs as a first line of therapy, and the remaining 10% of patients received glatiramer acetate.

The Basis for IFN-β Treatments in MS

IFN-β has a wide range of activities. Although the mechanisms of IFN-β action specific to MS are not completely elucidated, there is evidence that IFN-β is involved in a number of immuno-modulatory processes, which are important for MS patho-physiology. Established functions associated with IFN-β in MS pathophysiology include:

- Inhibition of the activated lymphocytes and macrophages migration through the damaged BBB into the CNS.
- Direct stabilization of the BBB.

- Reduction of the immune complex levels.

- Inhibition of the expression of major histocompatibility complex (MHC) molecules on the surface of antigen-presenting cells.

- Activation and stimulation of NK cell proliferation. NK cells may be one of the major immunological targets of IFN-β based treatment.

- Restoration of suppressor T cell functions.

- Marked reduction of pro-inflammatory cytokine release and an inhibition of their action.

References

Wagstaff and Goa, 1998; Yong et al, 1998; Perini et al, 2000; Hartrich et al, 2003; Feldman and Steinman.

Standard Tests for the Assessment of Treatment Efficacy

For the assessment of status of MS patients, and the assessment of treatment efficacy, the clinical and paraclinical tests—the rate of relapses, the score of disability, and a number of the brain lesions—are utilized. Clinical trials use a combination of short-term and long-term outcome measures, each of them being either clinical outcomes or MRI findings.

- In the most clinical trials, relapse is defined according to Schumacher and co-authors as the appearance of new symptoms or worsening of old symptoms attributable to MS, accompanied by new neurological abnormalities or focal neurological dysfunction lasting at least 24 hours, and preceded by the stability or improvement for at least 30 days. Usually in clinical trials, relapse-related measurements consist of several parameters such as the annual relapse rate, severity of relapse, and the percentage of the relapse-free subjects.

- The most important therapeutic aim of any MS-modifying treatment is to prevent or postpone long-term disability. The most commonly utilized tool for the disability assessment is the Kurtke Expanded Disability Status Scale (EDSS). The EDSS scores range from 0 to 10.0, where 0 means no disability and 10.0 means lethality due to MS. Disability progression is quantified as a persistent increase of one or more EDSS data points confirmed during two subsequent evaluations separated by 90 days. Recently, a modified disability test, the Integrated Disability Status Scale (IDSS), has been utilized in some clinical trials. The IDSS score quantifies both temporary and unremitting disabilities during the study period.

- MRI measures include assessment of the total extent of demyelination and the activity of process. The total extent of demyelination expressed as the cumulative area of lesions and referred to as the "burden of disease." The disease activity expressed as the number of new and enhancing lesions. It is well known that virtually all patients with MS experience ongoing disease activity and CNS damage even when no obvious clinical deterioration is observed. New brain lesions detected by MRI are not always associated with immediate clinical relapses. Sometimes, accelerated brain atrophy is present at the early and mild stages in relapse-free patients with stable EDSS scores. It was suggested that as many as 9 out of 10 active lesions are located in clinically silent areas of the brain and do not produce symptoms. This suggestion may explain the complexity of the relationship between disease activity detected on MRI and clinical status.

Nevertheless, MRI provides a dynamic measure of the MS brain pathology progression.

• There are no standardized criteria for MS treatment failure. For practical reasons, a distinction between primary and secondary treatment failure would be helpful. The primary failure pertains to the patients whose disease course is not influenced by the treatment at all. The secondary failure is defined as an initial stabilization of the disease with later recurrence of relapses or continuous progression of disability. A treatment of patients with RRMS is considered a failure if relapse rate remains constant or even increases, or disability continues to progress. In fact, the Quality Standard Subcommittee of the American Academy of Neurology proposed to consider a steady progression of disability for six months as a criterion for stopping the treatment with Betaseron.

References

Schumacher et al, 1965; Kurtke, 1983; INFB MS Study Group, 1993; Report of the Quality Standard Subcommittee of the America Academy of Neurology, 1994; Li et al, 1999; Rieckman et al, 1999; Goodin, 2001; Prisms Study Group and the University British Columbia MS/MRI Analysis Group, 2001; Kalkers et al, 2002; Miller et al, 2003.

Experimental Tests for the Assessment of Treatment Efficacy

Although IFN-β is effective in reducing the number of relapses and disease activity in MS patients, IFN-β-based drugs produce no benefits in almost one-half of all patients (non-responders). It is well known that during IFN-β therapy presence of neutralizing antibodies reduces or even abolishes the drug bioavailability. It is very important to develop sensitive and cost-effective laboratory methods for the early identification of non-responders. Several laboratory assays are currently under the development and, hopefully, they will provide a new insight into the MS biochemical pathology, and could be used as treatment outcome measures.

- Alteration of interactions between the endothelial cells and extracellular matrix (ECM) are critical events leading to the BBB damage. ECM integrity is mainly maintained by a dynamic balance between the synthesis and proteolysis of its components—matrix metalloproteinases (MMPs) and tissue inhibitors of metalloproteinases (TIMPs). Upregulation of MMPs is associated with endothelial injury leading to leukocyte passage through the BBB, leukocyte activation within the CNS, and direct degradation of myelin components. An imbalance between MMPs and TIPMs with over-expression of MMPs was detected in MS patients. MMPs presence and enzymatic function in the CSF of MS patients correlate with an increased clinical disease activity. MMP-mediated effects are currently recognized as the critical for the MS pathogenesis and clinical course. It was shown that after IFN-β therapy MMP-9 levels significantly decreased and TIMP-2 levels increased in comparison to values before the treatment. Published results (see Khuth et al.; Sobel; Waubant et al.; Giannelli et al.) suggested that MMP levels may be a useful index for the assessment of IFN-β efficacy.

- MxA is an antiviral protein. MxA intracellular expression is induced by viruses or IFNs. The studies demonstrated a correlation between MxA levels and MS clinical parameters. MxA levels were significantly lower during relapses and higher in the patients without relapses. In the stable IFN-β-treated patients, MxA levels were significantly increased compared to the patients without treatment (natural history control). Deisenhammer et al., Kracke et al., Bertolottoo et al., Vallittu et al., Bertolottoo et al., and Baranzini with co-authors suggested that the quantitative evaluation of MxA levels in the lysed blood cells could be one of the most appropriate markers for measuring a biological activity of exogenous IFN-β and a meaningful discriminating marker of the treatment efficacy.

- IFN-β induces the expression of the protein neopterin. Casoni and co-workers detected a significant decrease in neopterin levels in the patients with MS progression and suggested that neopterin may be considered as a useful marker for the responsiveness to IFN-β treatment.

- Gonen and co-authors showed that the patients with a marked reduction of N-acetylaspartate (NAA), which is consistent with a greater axon damage, have more severe disease and worse prognosis. Therefore, it was suggested that levels of NAA may be used as a prognostic sign and indicator of positive response to the treatment.

References

Deisenhammer et al. 2000; Kracke et al, 2000; Bertolotto et al, 2001; Khuth et al, 2001; Sobel, 2001; Waubant et al, 2001; Giannelli et al, 2002; Gonen et al, 2002; Vallittu et al, 2002; Bertolotto et al, 2003; Casoni et al, 2004; Baranzini et al, 2005.

Treatment Recommendations

Up-to-date, even after publication of the International Consensus Statement and several National Consensus Statements and Recommendations (the American Academy of Neurology, the Executive Committee of the Medical Advisory Board of the National MS Society, the Canadian MS Clinic Network, the

Australian-German-Swiss MS Therapy Consensus Group), there is
no single established guideline regarding the treatment of MS. All
recommendations include similar major points, emphasizing the
importance of early treatment and suggesting that the therapy
should be continued indefinitely, unless a clear lack of benefit or
intolerable side effects occur. The recommendations acknowledge
the effectiveness of IFN-β treatment, but do not give strict
guidelines to determine which of the current FDA approved
IFN-β drugs offers the greatest benefit, what the relative merits of
each immunomodulator are, and what agent is the optimal choice
for an individual or a group of MS patients.

References

Oger and Freedman, 1999; Freedman et al, 2002; Goodin et al, 2002.

<u>Key Points:</u>

1. The multifactorial pathophysiology of MS has fundamental
   implications for the development of appropriate therapies.
2. IFN-β-based drugs now are the first line of MS treatment.
   They normalize the immune status and regulate MS-
   associated excessive immune responses without resorting to
   a generalized immunosuppression.

The pharmacological treatment of MS is a young market, which
started after the FDA approval of the IFN-β drug, Betaseron.
Although IFN-β-based drugs and Copaxane improve the
manifestations of the natural course of MS, they do not provide a
cure for the disease. Intense research is underway for the
development of new drugs that are directly related to the
pathophysiology of MS. One of them is the monoclonal antibody
against α_4 integrins. The next chapter describes the structure and
functions of integrins.

Chapter 2

Integrins—Cell Adhesion Molecules

The ability of cells to adhere to other cells and to the extracellular matrix (ECM) plays a critical role in numerous processes. The cells adhere to other cells and to the ECM via cell adhesion molecules (CAMs), which create the cell-to-cell and cell-to-matrix networks, maintain the cell junctions, and define the degree of cell contact with microenvironment.

CAMs are cell-selective surface proteins, which are classified into several groups according to their molecular structure. These groups include the integrin family, the cadherins, the immunoglobulin superfamily, and the selectins. The integrins are important group of CAMs known to participate in all types of cell adhesion. The rest of this section describes the structure and functions of the integrin family in detail.

Structure

Integrins are transmembrane heterodimeric proteins composed of two non-covalently linked α- and β-chains. Integrin chains consist of extracellular domains, transmembrane domains, and cytoplasmic domains. For mammals, eighteen different types of α-subunits and eight β-subunits have been described. Based on the β-chains, integrins are classified into eight subfamilies, β_1 through β_8 integrin subfamilies; β_1 integrins are the largest subfamily of integrins. In each subfamily, an individual β-chain may be associated with several types of α-chains. Currently, 21 different integrin combinations have been recognized, which are referred to by their combinations of α- and β-chains. The combinations of α- and β-chains define integrin-ligand binding specificity.

Ligands

The cell-to-cell and cell-to-matrix networks are formed by binding of integrins with the set of cellular and ECM ligands specific to the integrins. For example, $\alpha_1\beta_2$ integrin interacts with the intracellular adhesion molecules 1 and 2 (ICAM-1 and ICAM-2); $\alpha_4\beta_7$ integrin binds to the mucosal adhesive cell adhesion molecule 1 (MadCAM-1); and $\alpha_4\beta_1$ integrin binds to the vascular cell adhesion molecule 1 (VCAM-1). Most integrins from the β subfamily recognize several ECM glycoproteins—such as fibronectin, vitronectin, laminin, and collagen—as ligands.

Functions

Initially integrins were considered important mostly for the immune responses due to their adhesion properties providing immune cell migration to the sites of inflammation. However, numerous recent studies indicated that integrins participate in a variety of processes on the systemic, tissue, cell, and molecular levels that include homeostasis and hematopoiesis; embryogenesis and fertility; establishment and maintenance of tissue architecture; coordination of apoptosis, survival, growth, and shape of cells; cell proliferation and differentiation; cell cycle division; and gene expression.

A broad range of specialized functions of the integrin family depend on the ability of integrins to serve not only as adhesion receptors but also as bidirectional signaling molecules, which can transduce biochemical stimuli within the effector cells.

Bidirectional signaling mechanism

Integrins are receptors that can transmit signals in both directions through the cell membrane. The ability of integrins to bind ligands is coordinated from the inside of the integrin-bearing cells (inside-out signaling), while the firm integrin-ligand binding elicits signals that are transmitted back into the cells and modulate intracellular processes (outside-in signaling). The integrin regulation represents a closed circle with intersecting link between the inside-out and outside-in signaling mechanisms (**Figure1**).

The majority of integrins are expressed on the surface of resting mature cells in an inactive state with low affinity to ligands. After activation of the resting cells by external or internal stimuli, intracellular signals regulate the conformational changes of the integrin ligand-binding sites located on the extracellular integrin domains. These conformational changes switch intergins into an adhesion-competent state with high affinity of integrins to its cellular and ECM ligands. The mechanism of integrin modification, which is controlled by the intracellular signals from within the activated cells, has been termed "inside-out" signaling.

Integrin-ligand binding, in turn, elicits signals back to the cells, which activates the cascade of intracellular reactions. The mechanism of regulation of intracellular processes, which is controlled by the signals generated by ligand-occupied integrin extracellular domains, has been called "outside-in" signaling. The outside-in signaling mechanism begins with the adhesion of

32   Integrins & Anti-integrins

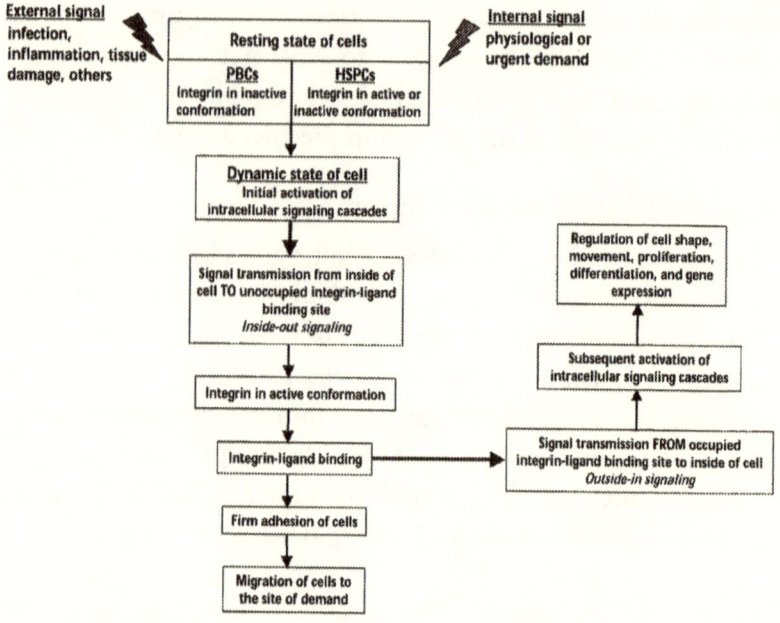

Figure 1. Scheme of integrin bidirectional signaling in hematopoietic and mature blood cells

Integrins serve as linkers for the bidirectional circuit of cell's inside-out and outside-in signaling. Peripheral blood cells (PBCs) are activated by external extracellular stimuli such as infection, inflammation, tissue damage, or other; hematopoietic stem and progenitor cells (HSPCs) are activated mostly by physiological or urgent (for instance, blood loss) demands. Stimulus-mediated initial cell activation induces the cascade of intracellular reactions that resulting in change of inactive integrin conformation to active adhesion conformation on the extracellular intergin domains—inside-out signaling.

Adhesive-competent integrins bind to their ligands. Signal transduction from the integrin-ligand binding site causes two cell phenomena: firm adhesion of the integrin-bearing cells to other cells or matrix following by the migration of cells to the site on demand; and feed-back activation of intracellular regulatory pathways following by changes in the cell metabolism, shape and movement, proliferation, differentiation, and gene expression—outside-in signaling.

integrins to their ligands and modulates many intracellular pathways, which influence vital cellular processes including cell proliferation, differentiation, and gene expression.

Although the molecular basis for integrin-mediated bidirectional signaling and detailed signal transduction pathways have not been completely elucidated, the latest crystallographic, electron microscopic, and biochemical studies have provided new data based on which the hypotheses of the integrin functioning have been proposed. All hypotheses suggested that the inside-out and outside-in mechanisms both involve the bidirectional integrin interdomain communications through the transmembrane signals.

The experimental evidence strongly supports models of the inside-out signaling, which suggest that the intracellular signals from the activated integrin-bearing cells firstly stimulate conformational changes in the integrin cytoplasmic domains. The conformational changes of the integrin cytoplasmic domains propagate through its transmembrane domains to the extracellular domains. The activation of the ligand-binding site of integrins is caused by tertiary and quaternary conformational changes in the extracellular domains, which transform integrin from a low to high affinity ligand-binding receptor. Models of the outside-in signaling, on the other hand, also suggest that the outside-in mechanism must be dependent upon the conformational state of the integrin cytoplasmic domains. After integrin-ligand binding, the ligand-occupied sites transmit signals across the cell plasma membrane to the cytoplasmic domains. Signal transmission induces structural rearrangements in the integrin cytoplasmic domains, which initiates the cascade of intracellular reactions. Therefore, the integrin cytoplasmic domains are the center of bidirectional signaling machinery and hence play a critical role in the integrin functions.

Note: The α_4 integrin, particularly $\alpha_4\beta_1$ integrin, is also known as a very late antigen-4 (VLA-4). This term was coined because the α_4 integrin was first identified on a T cell surface at a late stage of T cell maturation. However, later experiments showed expression of this protein on the surface of various cells. Therefore, the term "α_4 integrin" will be used here in most cases instead of the historical name "VLA-4".

Refernces

Hynes, 1987; Nojima et al, 1992; Freedman et al, 1993; Salomon et al, 1994; McGilvray et al, 1997; Bronson et al, 1999; Giancotti and Ruolahti, 1999; Papadaki, 1999; Grabovsky et al, 2000; Feigelson et al, 2001; Hynes, 2002; Schwartz and Ginsberg, 2002; Xiong et al, 2003; Calvete, 2004; Kaplan et al, 2005.

Key Points:

1. Integrins are bidirectional signaling molecules that regulate a broad range of vital cell processes.

Topographical and Functional Diversity of α_4 Integrins

Immune System

A plethora of data demonstrates that α_4 integrins are expressed by multiple mature cells in the blood circulation, including T and B lymphocytes, NK cells, monocytes, and granulocytes (eosonophils and basophils, but not neutrophils) (**Figure 2**).

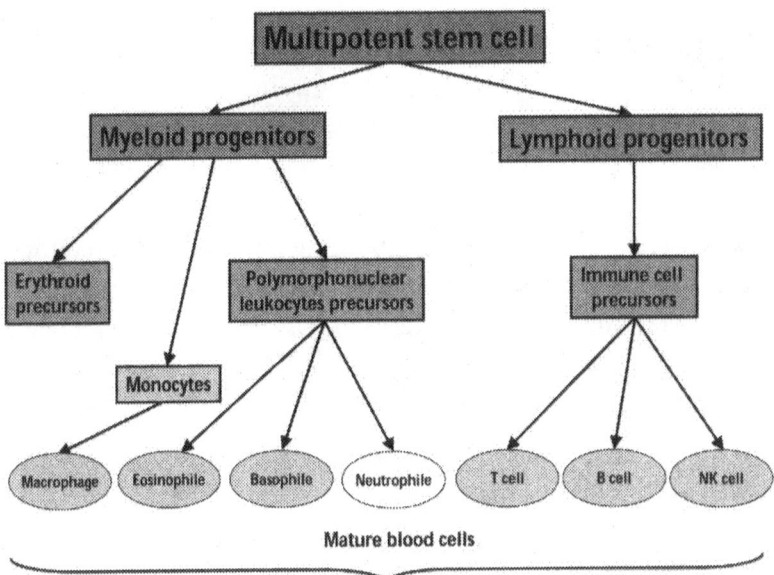

Figure 2. Expression of α_4 integrins on the surface of hematopoietic and mature blood cells

α_4 integrins are expressed on the surface of multipotent stem cell, all classes of the progenitor cells of erythroid, myeloid, and lymphoid lineages, and on the surface of mature blood cells—T cells, B cells, NK cells, granulocytes (excluding neutrophils), and macrophages.

As a component of immune cells, integrins participate in both the inside-out and outside-in signaling mechanisms. The inside-out mechanism provides recruitment of leukocytes during the inflammatory response. Integrins are involved in the migration of immune cells from the vessels into the tissues, in the uptake of immune cells at the sites of inflammation or tissue damage, and in the recirculation of immune cells throughout the lymphoid organs. The most important integrins, which participate in T cell migration, are two α_4 integrins—$\alpha_4\beta_1$ and $\alpha_4\beta_7$. The α_4 subunit is shared by both proteins, and the β subunit varieties mostly determine the integrin-ligand binding specificity. Therefore, adhesion molecules serve not only to facilitate the migration of lymphocytes but also contribute to the tissue-specific lymphocyte trafficking. The leukocyte trafficking is a complex process depending on the presence of specific humoral factors at the site of inflammation, the particular leukocyte subset, the state of vascular endothelium, and the activation of integrins.

Under the baseline conditions, integrins have a relatively inactive conformation keeping the leukocytes in a non-adhesive form. The activation of immune cells under infection and inflammation generates signals inside of the leukocytes resulting in conformational changes of the integrins, which dramatically increase the adhesiveness of leukocytes to the vessel endothelial receptors. Ligand-competent α_4 integrin mediates firm adhesion of rolling leukocytes to the endothelium via integrin-ligand binding with following transendothelial migration to the site of inflammation.

Adhesion interaction is required for a broad spectrum of immune cell functions, including T cell-mediated killing, helper T cell activity, NK cell-mediated action, B cell response, and antibody-dependent cytotoxicity mediated by monocytes. Leukocyte trafficking is essential to the immune host defense. Under some conditions, leukocyte-endothelial interactions may cause an excessive leukocyte accumulation and contribute to the tissue damage and the development of diseases (for instance, MS, rheumatois arthritis, Crohn's disease). However, partial or total absence of the leukocyte adhesion results in defective recruitment of leukocytes to the inflammation sites, decreased the immune functions, and impaired the defense against infections. Thus, prolonged blockage of the leukocyte adhesion leads to the risk of patient susceptibility to bacterial and virus infections.

References

Hemler, 1990; Nojima YD et al, 1992; Freedman et al, 1993; Eisenmann and Kim, 1994; Hunter, 1995; Sato et al, 1995; McGilvray et al, 1997; Papadaki, 1999; Grabovsky et al, 2000; Zhan et al, 2000; Feigelson et al, 2001; Van Assche and Rutgeerts, 2002.

Key Points:

1. Integrins of $\alpha_4\beta$ family are expressed on the surface of almost all mature immune cells.

2. Decreased expression or prolonged blockage of $\alpha_4\beta$ integrin functionality may lower the immune resistance to bacterial and viral infections.

Hematopoietic Stem and Progenitor Cells

The majority of hematopoietic stem and progenitor cells (HSPCs) express $\alpha_4\beta$ integrins. HSPCs at various stages of the development and differentiation display various patterns of integrin expression. Mature cells exhibit a diminished expression of $\alpha_4\beta$ integrins in contrast to the HSPCs, which express $\alpha_4\beta$ integrins in a constitutively active state. The abundance of $\alpha_4\beta$ integrins on the surface of immature hematopoietic cells suggests that these integrins are involved in the trafficking and activity of all classes of hematopoietic cells—erythroid, myeloid, and mixed hematopoietic progenitors (**Figure 2**).

In a steady state, immature proliferating hematopoietic cells at various stages of differentiation are confined within the specialized bone marrow (BM) "niches," whereas terminally differentiated mature cells leave the BM and migrate into the blood. This dynamic process allows for an accelerated expansion and migration of the mature blood elements out of the BM to meet physiological or urgent demands. In addition, a small proportion of morphologically unrecognizable primitive stem cells regularly escape from the BM into the circulation. Therefore, it may be stated that blood cells belonging to the two extremes of the developmental spectrum are continuously circulating in the peripheral blood (PB).

Numerous experiments over the last years elucidated the effects of $\alpha_4\beta$ integrin interactions with their ligands in the BM. In the most of these studies, anti-$\alpha_4\beta$ integrin monoclonal antibodies were utilized to block integrin functions. Particularly remarkable experiments were performed on the animal models (baboons, macaques, and mice) and on the long-term cell cultures. In other set of experiments, α_4 integrin deficient chimeric mice and α_4 integrin ligand (VCAM-1)-deficient chimeric mice were studied.

Dr. Popoyannopoulou and co-workers as well as by other researchers, in sophisticated experiments which were performed over many years, have obtained strongly conclusive results elucidating the critical role of integrin $\alpha_4\beta_1$ in the HSPCs cell mobilization (recruitment of HSPCs from the BM into the blood) and revealing the molecular mechanisms of this process. The studies with baboons and macaques were the first to determine that intravenous injection of saturating amounts of anti-$\alpha_4\beta_1$ integrin antibody increased up to 200-fold the mobilization of all classes of progenitor cells. The accelerated progenitor cell mobilization after anti-$\alpha_4\beta_1$ integrin antibody treatment was confirmed by the experiments with mice, which have shown that following the injection of anti-$\alpha_4\beta_1$ integrin antibody the progenitor cells with a high self-renewal potential were recruited into the blood. Anti-$\alpha_4\beta_1$ integrin antibody inhibited binding of the adhesion molecules expressed on the surface of hematopoietic progenitors to their ligand VCAM-1 expressed by stromal cells.

Analysis of the molecular components involved in the mobilization process showed that the most significant antigenic and functional phenotype differences between mobilized progenitor cells and steady-state BM cells are the absence of cycling and a decreased expression of $\alpha_4\beta_1$ in the mobilized cells. Functional down-regulation of integrin is a common step at the final stages of transmigration of hematopoietic progenitor cells through the endothelial sinuses. This suggests that $\alpha_4\beta_1$ integrin plays an important role in the regulation of cell retention within the BM. It is important to emphasize that down-regulation of $\alpha_4\beta_1$ integrin by itself is not sufficient to induce the mobilization process in all cells. Additional complex interactions and cooperative signaling with other pathways of migration may be necessary for the cells in order to exit the BM.

The role of integrin in the mobilization process was confirmed unequivocally by data with α_4 integrin-deficient and α_4 integrin ligand VCAM-1-deficient chimeric mice. The deletion of the

α_4 integrin gene resulted in an efflux of hematopoietic progenitor cells from the BM into the circulation, which continued for as long as the mice were tested (over 12 months). The egress of the progenitor cells into the circulation sustained in α_4 integrin-deficient mice for over 12 months strongly suggests that α_4 integrin in the primitive multipotent progenitor cells was responsible for the release of all progenitors found in the circulation. No compensatory mechanisms emerged to correct this phenotype. Additionally, mice with conditional ablation of VCAM-1 had elevated levels of immature B cell in the PB and reduced mature lymphocytes number in the BM. The fact that $\alpha_4\beta_1$ and VCAM-1 acted as mediators of HSPCs mobilization was confirmed by in vitro experiments, where antibodies against $\alpha_4\beta_1$ or VCAM-1 molecules prevented stem cells from the binding to fibronectin coated dishes.

Besides involvement in the immobilization process, $\alpha_4\beta_1$ integrin participates in hematopoiesis, specifically, in the differentiation and proliferation of HSPCs in the BM. Set of experiments with α_4 or β_1 integrin-deficient mice and experiments with pregnant mice injected with anti-α_4 integrin antibodies showed that within the lymphoid germinal centers $\alpha_4\beta_1$ plays a key role during HSPC ontogeny. It was shown that $\alpha_4\beta_1$ deficiency leads to the impairment of progenitor cell terminal differentiation. Furthermore, the deletion of either the α_4 or β_1 integrin gene caused embryonic lethality from non-hematological defects in mice.

Integrin mediates stable contact between HSPCs and stromal cells and between HSPCs and extracellular matrix components. The results of numerous experiments provided direct evidence of $\alpha_4\beta_1$ integrin involvement in hematopoiesis and demonstrated that a perturabation of α_4 integrin functioning prevented the development of erythroid, lymphoid, and myeloid progenitors into mature cells.

References

Papayannopoulou and Nakamoto, 1993; Freedman et al, 1993; Strobel et al, 1997; Frenette et al, 1998; Quesenberry et al, 1998; Vermeulen et al, 1998; Papadaki, 1999; Grabovsky et al, 2000; Kronrwett et al, 2000; Papayannopoulou, 2000; Gazitt et al, 2001; Koni et al, 2001; Scott et al, 2003; Boning, 2004; Papayannopoulou, 2004; Ulyanova et al, 2005.

Key Points:

1. Homeostasis, blood formation, and functions of HSPC depend on their relationship with adjacent cells and extracellular matrix within the BM.

2. Cell-to-cell and cell-to-extracellular matrix interactions within the BM are mediated by $\alpha_4\beta_1$ integrin-ligand binding. These interactions are critical to the regulation of HSPC differentiation and proliferation; the migration and anchoring of HSPCs within the BM; the retention of immature blood cells within the BM; and the transmigration of mature blood cells into the bloodstream.

3. Perturbation of α_4 integrin functioning may block the development of erythroid, lymphoid, and myeloid progenitors into mature cells.

4. Alteration in the adhesive interactions between HSPCs and BM stromal cells as well as between HSPCs and extracellular matrix may lead to the continuous proliferation and abnormal circulation of progenitors.

5. Experiments showed that the injection of anti-$\alpha_4\beta_1$ antibodies mobilize a wide range of progenitor cells including progenitors with a high proliferative potential. It is crucial to note that an increased mobilization of immature progenitors from the BM to the peripheral blood may have dire consequences.

Germ Cells, Fertilization, and Embryonic Development

Fertilization and embryonic development are the processes with highly ordered sequence of the events. Sperm-egg fusion initiates the embryonic development. The responses following sperm-egg attachment are strikingly similar to the responses of lymphocytes following the activation by antigen presenting cells. Similar to the adhesion of the somatic cells, sperm-egg adhesion and fusion involve combined receptor-ligand interactions, which are initiated primarily by proteins of the integrin family.

Various lines of experiments demonstrated that integrin molecules are present on the surface of germ cells and mature mammalian gametes: integrin subunits α_2, α_4, β_1, β_2, and β_7 were observed to localize on the human oocytes; ejaculated spermatozoa contain the mRNA transcripts of several β_1 integrins and matrix proteins. Embryonic cells express developmentally regulated complex repertoire of adhesion receptors. A variety of integrin subunits have been detected on the human embryos. Integrin $\alpha_4\beta_1$ was shown to localize at all stages of embryonic development except for during blastocyst outgrowth (**Figure 3**). The highest level of $\alpha_4\beta_1$ expression was detected at the earliest stage of embryogenesis. It is logical to suppose that simultaneous differential expression of various integrins at the early stages of embryonic development may

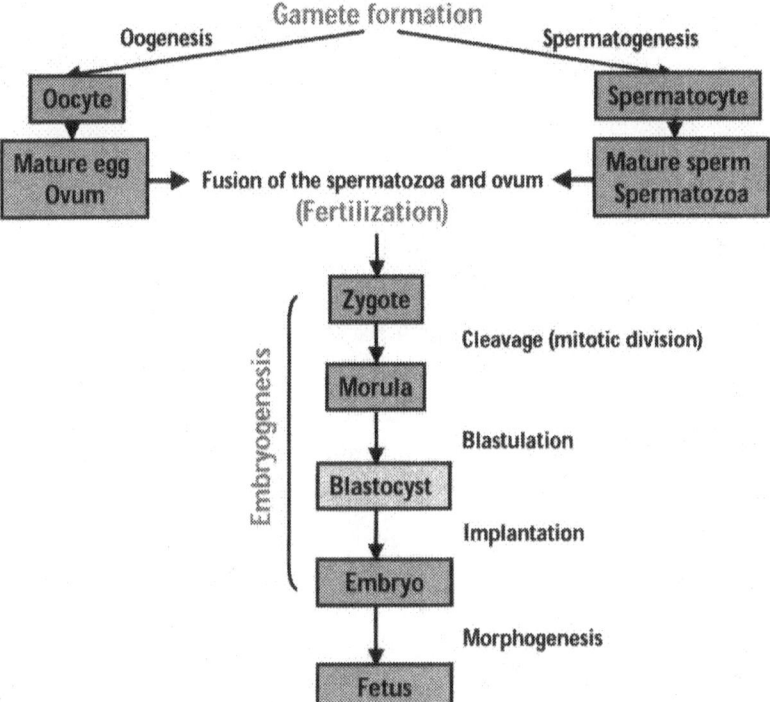

Figure 3. Expression of α_4 integrins on the surface of germ, gamete, and embryonic cells

α_4 integrins are expressed on the surface of germ cells during gamete formation, gamete cells, and embryonic cells. $\alpha_4\beta_1$ integrin was not detected on the blastocyst stage of development.

contribute to the fine modulation of cellular interactions during the adhesive processes. Dynamic changes of integrin expression may be important for continuing embryonic development, survival, and interactions between the embryo and the maternal environment, which is essential for establishing maternal tolerance to the fetal allograft. Apart from the adhesion function, integrin-ligand interactions mediate signals in the embryonic cells inducing changes in tyrosine phosphorylation of oocyte proteins, substantial cytoskeletal reorganization, and gene expression. Additionally, positive correlation was identified between the expression of integrins α_4, α_5, and α_6 and successful fertilization. Compared to the normal sperm, sperm of infertile patients exhibits significantly decreased expression of these adhesion molecules. Defects in the sperm-egg fusion may account for some forms of infertility.

Experimental evidence proved that the integrin-mediated cell-to-cell interactions and binding to ECM ligands (fibronectin, laminin, and vitronectin) play a crucial role in the major reproductive processes—spermatogenesis, sperm-egg fusion, and embryogenesis.

References

Bronson and Fusi, 1990; Lathrop et al, 1990; Wilhams et al, 1992; Burrows et al, 1993; Schaller and Parson, 1993; Schaller et al, 1993; Tarone et al, 1993; Almeida et al, 1995; Klentzeris et al, 1995; Rohwedder et al, 1996; Bronson et al, 1999; Lu et al, 2002.

Key points:

1. Integrins play the crucial role in the process of reproduction. They are involved in the major reproductive processes: spermatogenesis, oogenesis, fertilization, and embryogenesis.

2. The impairment of the integrin functions may result in some forms of infertility and embryological defects.

Other Physiological Processes

In addition to the major impact of the $\alpha_4\beta_1$ integrin on immunity, hematopoiesis, and reproduction it was demonstrated that other physiological processes are influenced by the adhesion molecules: integrin $\alpha_4\beta_1$ plays an important role in the contact-mediated microglial expression of inflammatory cytokines and modulates the genesis of vascular myogenic tone.

References

Davis et al, 2001; Dasgupta, 2003

Message to the Readers

Integrin $\alpha_4\beta_1$ is expressed on the surface of almost all immune cells. It is required for a broad spectrum of innate and adaptive immune cell functions. Decreased leukocyte adhesion leads to deficiency of the immune responses, which in turn reduces resistance against infections. Thus, prolonged treatment with anti-$\alpha_4\beta_1$ integrin antibodies renders patients susceptible to bacterial and viral infections.

Integrin $\alpha_4\beta_1$ is expressed on the surface of the blood stem and progenitor cells and determines their adhesive properties. Cell adhesion plays a crucial role in the regulation of progenitor cell proliferation and differentiation, as well as in the retention of immature blood cells within the bone marrow and the release of mature cells into the bloodstream. Chronic reduction of $\alpha_4\beta_1$ functionality may lead to abnormalities in the hematopoietic system. Long-term blockage of hematopoietic progenitor cell adhesion with anti-$\alpha_4\beta_1$ integrin antibodies raises the theoretical but, nevertheless, real concerns for an increased risk of hematological malignancies.

Integrin $\alpha_4\beta_1$ is expressed on the surface of the germ cells and plays a crucial role in the spermatogenesis, fertilization, and embryonic development. Diminished expression of $\alpha_4\beta_1$ integrin or inhibition of its functions by anti-$\alpha_4\beta_1$ integrin antibodies may lead to the alterations in spermatogenesis, reduced fertility, and defects in embryonic development.

The molecular mechanisms responsible for the cellular adhesion are known to be complex, well orchestrated, and sophisticatedly controlled processes. The $\alpha_4\beta_1$ integrin is involved in a remarkably broad range of biological processes. Alterations in the expression of cell adhesion molecules or blockage of their function may lead to the dramatic remodeling of these processes, and therefore, the loss of the adhesion interactions may result in severe pathological consequences. It is essential to recognize that integrin-antagonist therapy, i.e., thereapy with anti-α_4 integrin monoclonal antibodies, is a double-edged sword, which must be handled with extreme caution.

Following chapter discribes the general principles of the monoclonal antibody development and presents detail analysis of the clinical trials with anti-α_4 integrin monoclonal antibodies, natalizimab (Tysabri).

Chapter 3

Monoclonal Antibody Therapy

In 1975, Kohler and Milstein reported their successful collaboration on the technique of monoclonal antibody (Mab) production based on the ingenious new principle. The idea was to develop the specific antibody from a single clone of B lymphocytes to be produced by immortal cells in the cell culture. For this purpose, a mouse is injected with the antigen. After the appropriate length of time for antibody production, the animal's spleen is removed. The spleen is rich in B lymphocytes and plasma cells, and each cell is genetically programmed to produce an antibody of single specificity. Plasma cells, however, do not grow in cell culture. For that reason, a plasma cell producing a selected antibody is fused with an immortal myeloma cell. Thus a fused cell—hybridoma—is created, which endlessly replicates in the cell culture. The hybridoma has the characteristics of the both parental cells: it endlessly and rapidly grows in the cell culture as a myeloma cell and produces a large amount of specific antibody as a plasma cell.

The creation of unique hybridoma technology was one of the major breakthroughs in immunology heralding a new era in immunological research and clinical development. The scientific and medical community quickly recognized the significance of this discovery, and in 1984 the two scientists were awarded the Nobel Prize for Medicine.

Structure

Mabs are homogenous sets of immunoglobulins (Igs), mostly IgG, with the same basic structure as all IgG class antibodies, composed of two heavy chains and two light chains. The heavy chains form a fused "Y" structure; the two light chains run parallel to the open portion of the heavy chains. The four chains are covalently linked by disulfide bonds. Each chain is made up of variable (V) regions and constant (C) regions (also called domains). Heavy chain has one variable (V_H) and three constant (C_H) regions, and light chain has one variable (V_L) and one constant unit (C_L) (**Figure 4**).

Antigen-binding sites are formed on the V_L and V_H regions, they are called Fab fragment (antibody binding fragment). Within each Fab fragment there are the hypervariable regions unique to each IgG molecule. The hypervariable region, or complementary-determining region (CDR), constitutes the site where antigen recognition and binding occurs.

The C regions of IgG molecules are made of homologous sequences and have exactly the same primary structure as all other IgG chains of the same isotope. The C_H region at the terminal end of the IgG molecule is called the Fc fragment (**Figure 5**).

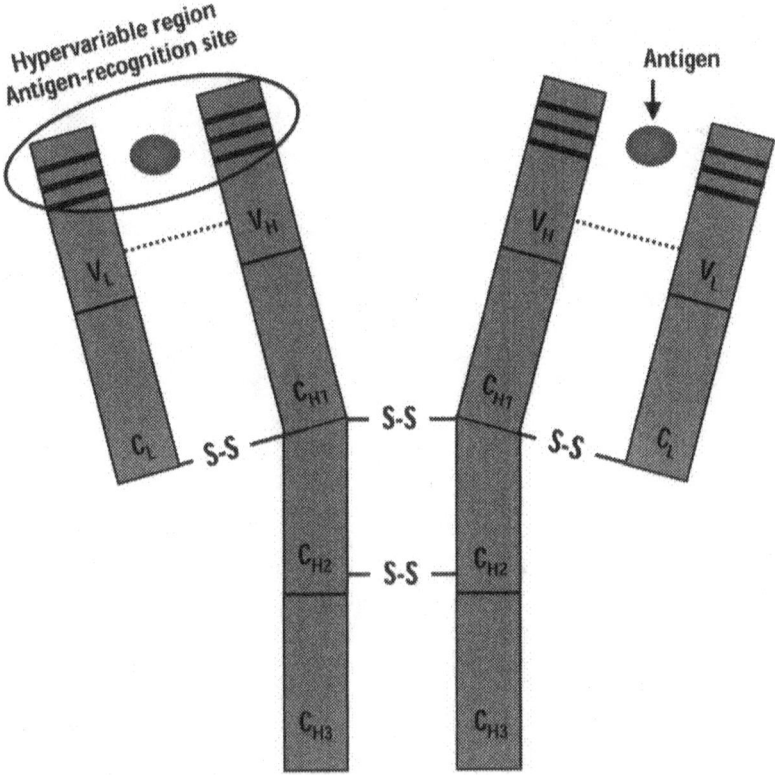

Figure 4. Schematic drawing of IgG antibody structure
JgG antibodies are composed of the two heavy and two light chains covalently linked by disulfide bonds. Each chain consists of the variable (V) and constant (C) regions. Within each V region there is hypervariable region, complementary-determining region, which constitutes the antigen-binding site.

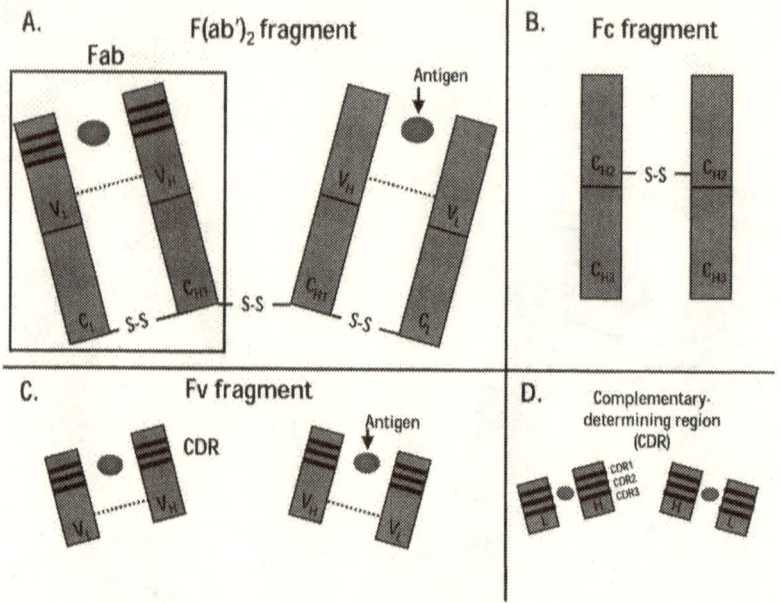

Figure 5. Schematic drawing of antibody's fragments

A. F(ab)₂ and Fab fragments are composed of the C_H, C_L, V_H, and V_L domains. C_H and C_L domains are bound by disulfide bonds; V_H and V_L domains are connected by non-covalent bonds. $F(ab)_2$ fragment (MW ~ 100 kDa; ~ 66% of the whole antibody's MW) contains two antigen-binding sites. Fab fragment (MW ~ 50 kDa; ~ 30% of the whole antibody's MW) contains a single antigen-binding site.

B. Fc fragment is the constant region of antibody (MW ~ 50 kDa; ~ 30% of the whole antibody's MW). Fc fragment mediates several antibody functions including compliment binding, which triggers elimination of antigen/antibody complexes.

C. Fv fragment is composed of the entire V_H and V_L domains, which connected by non-covalent bonds. Fv fragment (MW ~ 50 kDa; ~ 30% of the whole antibody's MW) contains two antigen-binding sites. Single-chain Fv (scFv) fragment (MW ~ 25 kDa; ~ 15% of the whole antibody's MW) contains a single antigen-binding site.

D. Complementary-determining region (CDR) is located at the N-terminal of the antibody. This region is an extremely variable in amino acid consequences and forms the unique antigen-recognition/binding site. CDR (MW ~ 17 kDa; ~ 10% of the whole antibody's MW) contains two antigen-binding sites. CDR (MW ~ 8.5 kDa; ~ 5% of the whole antibody's MW) contains a single antigen-binding site.

Types of Mabs

Antibody technology has made giant leaps forward since the first generation of murine Mabs. Throughout the 1990s, the utilization of the innovative recombinant DNA techniques has led to the production of chimeric Mabs, humanized Mabs, fully human Mabs, small Mab's fragmenst, and bispecific Mabs (**Figure 6**). New Mabs minimized the development of human anti-mouse antibodies production, decreased side effects, and enhanced the clinical efficacy of the therapeutic Mabs.

Murine Mabs

From the outset, scientists were very enthusiastic about the applications of Mabs. Mabs were incorporated into clinical practice in the early 1980s for therapeutic treatment and diagnostic purpose. Unfortunately, Mabs produced by the mouse hybridoma system originally developed by Kohler and Milstein did not live up to their great expectations for the first 20 years, with disappointing results regarding their use as therapeutic agents. At the same time, utilizing mouse Mabs for diagnostic purposes was very successful, and today the Mab technique is one of the most powerful diagnostic tools available for diagnosis of a wide range of diseases.

Therapeutic failure of the first generation of murine Mabs, which were developed for multiple therapeutic applications, was mostly due to the murine Mab's immunogenecity. The mouse IgG with molecular weight (MW) of about 150 kDa triggers the development of human anti-mouse antibodies (HAMAs). Human antibodies against variable regions of the mouse Mabs, anti-idiotypic antibodies, bind to the antigen-recognizing sites on Mabs and abolish therapeutic efficacy of the originally active drugs. Because

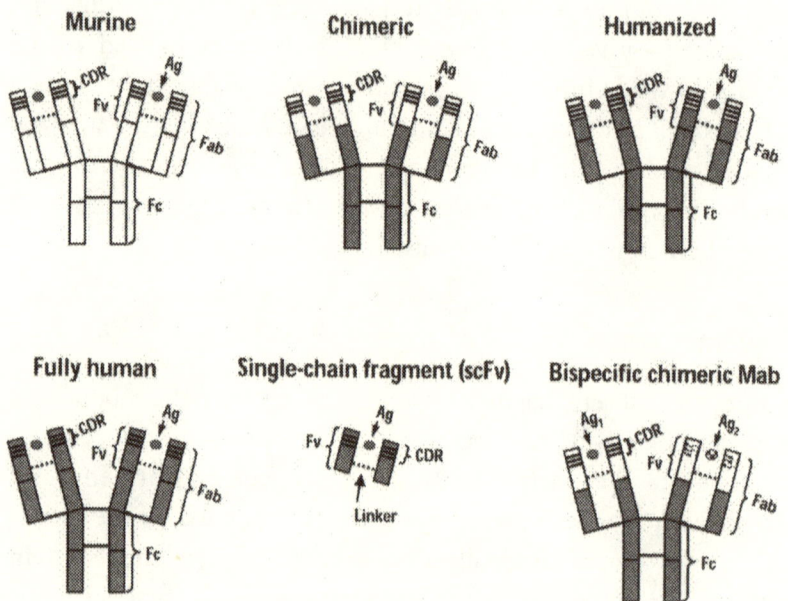

Figure 6. Schematic drawing of different types of monoclonal antibodies
Murine Mabs contain 100% of mouse amino acid consequences.
Chimeric Mabs is composed of mouse Fv region and human constant (C) region; they contain ~ 25–30% of mouse amino acid consequences and ~ 70% of human amino acid consequences.
Humanized Mabs is composed of mouse CDR region and human IgG framework; they contain ~ 10% of mouse amino acid consequences and ~ 90% of human amino acid consequences.
Fully human Mabs contain 100% of human amino acid consequences.
Small Mab fragment (scFv) is a construction in which the V_H and V_L chains of Fv domain are bound via a flexible peptide-linker.
Bispecific Mabs (chimeric Mab is presented as an example) has two different antigen-binding sites, and can, therefore, bind two different antigens.

repeated Mab administration provokes an anti-mouse immune response, typically, mouse Mabs can be used only once or twice before HAMA production decreases or even completely blocks their functions.

Using the same hybridoma technique, Mabs could be manufactured in the human cells, which would ameliorate the problem of HAMA production. However, few volunteers would agree to be immunized in order to elicit antibodies that they don't need. Nevertheless, attempts at human-derived Mab production have been made, but most of them were unsuccessful.

Serious problems with immunogenic reactions to murine-derived Mabs raised major doubts whether Mab utilization would ever emerge as a practical therapeutic option. But, the obvious potential applications of specific antibodies for treating cancer and immune disorders provided a substantial incentive for further achievements in Mab production technology.

Chimeric Mabs

Using genetic engineering, the second generation of Mabs containing both mouse and human sections, "chimeric" Mabs, were prepared. The chimeric antibodies combine the mouse IgG variable regions with the constant regions of a human IgG. Chimeric Mabs have decreased immunogenecity compared to the whole mouse Mabs because in the mouse-human hybrid Mabs only 25% to 30% of the total protein composition belongs to the mouse-derived components. Moreover, there are two more reasons why chimeric Mabs are more effective than mouse Mabs: because of the better pharmacokinetics, and because the human Fc fragment of chimeric Mabs activates the human complement system and cytotoxic cells for the destruction of the Mab-targeted cells more effectively than does mouse Fc fragment.

Humanized Mabs

Grafting of the mouse CDR into the human acceptor IgG framework was used for the development of third generation of Mabs, "humanized" antibodies. Humanized Mabs contain the mouse hypervariable regions of the Fab fragment, and the rest of antibody is entirely human IgG. The process of the monoclonal antibody "humanization" reduces the risk of HAMA occurrence compared to chimeric Mabs because humanized Mabs have in them only 5% to 10% of the mouse components versus 25%–30% in the chimeric Mabs. The immunogenecity is a problem during Mab treatments in general, and even using humanized Mabs it is still impossible to avoid HAMA production completely because the presence of some fraction of the mouse amino acids responsible for the antigen binding sites is inevitable.

In addition to the commonly utilized "humanization" method based on CDR grafting, some newly developed techniques permit the modification of any non-human Mabs in such a way that they resemble humanized Mabs. For example, Human Engineering™ is a proprietary technology, which uses unique algorithm based on the analysis of the conserved structure-function relationship among antibodies. The Human Engineering™ technology claims that antibodies preserve the antigen-binding structure and functions, greatly reduce or even eliminate the immunogenicity, and retain the important antibody functions such as complement-dependent cytolysis and antibody-dependent cellular cytotoxicity.

Fully Human Mabs

Modern alternative strategies allow for the selection of fully human Mabs, which may be considered to be the fourth generation of Mabs. Fully human Mabs can be produced directly from the natural or synthetic repertories including transgenic mice; antibody-display libraries; human B lymphocytes "immortalized" with Epstein-Barr virus, recombinant DNA techniques with the insertion of human genes coding for the variable Fab portion of antibodies into the genome of bacteriophages; and gene therapy procedures. Currently, several fully human Mabs are under clinical evaluation, and recent data clearly demonstrate their efficacy and safety.

Transgenic mice with the human antibody repertoire instead of the innate mouse immune inventory have been generated. This procedure involves utilizing mouse embryonic stem cells into which human antibody genes were inserted. Injection of a given antigen into these transgenic mice resulted in the development of fully human antibodies. Then the antibodies can be reproduced by the classic hybridoma method or, in order to increase affinity of antibodies, by *in vitro* display library and selection technologies. Both transgenic mice and display libraries techniques typically produce the human antibodies with binding affinity ranging from 10^{-7} to 10^{-9} M. Antibody affinity can be enhanced and optimized via site-specific mutagenesis, combinatorial mutagenesis of CDRs, or random mutagenesis of the entire genes.

Fully human Mabs are also being produced via recombinant DNA techniques by inserting the human genes coding for the variable Fab portion into the genome of bacteriophages. As the bacteriophages replicate, they display the Fab portion of the human antibodies on their surface. They are subsequently mixed with an antigen to select those bacteriophages that produce the complementary Fabs. The selected bacteriophage genome is then converted into plasmid, which subsequently produces soluble specific Fabs in bacteria.

The production of fully human Mabs utilizing a gene therapy protocol is a promising new approach. Experiments have shown that this method generated high and constant levels of antibody expression directly in the body. Moreover, a high concentration of antibodies can be achieved in close proximity to the target cells by implanting the recombinant Mab producers nearby. Although it was postulated that this approach could be highly beneficial and applicable to the production of Mabs *in vivo*, further studies are necessary to establish the most effective protocol, evaluate the gene/vehicle transfer safety, identify the most suitable transcriptional control elements, and compare the local versus systemic antibody production.

Small Mab's Fragments

Intact multivalent Mabs provide high specificity, high functional affinity, impressively long serum half-life, and their human Fc domain is important for cytotoxic effector functions. At the same time, utilization of intact Mabs is limited because their physical properties turn out to be inapplicable for certain purposes.

First, the large size of Mabs (150–200 kDa) makes it difficult for them to penetrate into the solid tumors. In fact, only 0.001%–0.01% of the injected Mab dose is incorporated into the solid tumors. Low penetration of Mabs into the solid tumors decreases their cancer cytotoxicity and dramatically increases side effects. As a result, solid tumors have often proven to be relatively resistant to the Mab-based therapies. Secondly, during radioimmunotherapy, prolonged circulation of intact Mabs leads to bone marrow exposure to the radioisotopes, which is associated with an unacceptable myelotoxicity. Additionally, even the Fc-induced effector functions are undesirable for particular clinical purposes.

Therefore, recombinant small Mab fragments of approximately 25 kDa were produced. New Mab formats, the Fab fragment and the single-chain Fv fragment (scFv), provide full antigen-binding specificity, rapidly penetrate into the solid tumors, and exhibit a negligible immunogenicity. Moreover, to increase the functional affinity, Fab and scFv fragments have been conjugated into dimeric, trimeric, or tetrameric structures via chemical or genetically engineered cross-links. The microbial expression of scFv is currently the favored method of the small fragment production.

Nevertheless, the practical use of small Mab fragments is still limited because the monovalent Mab fragments exhibit rapid blood clearance and short retention time in the targets, which necessitate their frequent delivery into the body. Mab-based gene-therapy circumvents the aforementioned limitations. Mab gene-therapy provides sustained concentration of soluble Mab fragments resulting in long-term anti-tumor efficacy and supports the expression of Mab domains on precise intracellular locations. The utilization of Mab-based gene therapy has a potential to revolutionize cancer therapy. Further improvements in this strategy require the design of *in vivo* selection systems to generate antibodies active in the specific cellular compartments as well as the development of suitable and safe gene-delivery systems.

Bispecific Mabs

In order to increase the efficacy of any type of Mabs, chimeric, humanized, or fully human, it is possible to conjugate them with a vast range of "killing" effector agents, creating so called bispecific Mabs. A bispecific Mab has two binding sites with different specificity and can, therefore, bind two different antigens. The concept of using bispecific Mabs for therapy is based on the selective recruitment of effector agents to a defined disease-related target structure. Antibodies are ideally suited as shuttles for bispecific molecules since they bind specifically and with high affinity to target antigens. The therapeutic mechanisms of bispecific Mabs include the delivery of effector molecules (radionuclides, toxins, drugs, pro-drugs, cytokines, and complement compounds), as well as the retargeting of effector cells (CTL, NK, macrophages, and granulocytes) and carrier systems (viral vectors for gene therapy).

Bispecific Mabs are powerful therapeutic agents covering various therapeutic strategies, including radioimmunotherapy, chemoimmunotherapy, immunotherapy, and gene therapy. Currently, the most applications of recombinant bispecific antibodies are for cancer therapy, and focus on the retargeting of effector cells of the immune system to the tumor cells. To date, a large number of target antigens are evaluated. Most of these are the tumor-associated antigens, which are over-expressed by the tumor cells. So far, the results are somewhat encouraging.

Mab's Market

During the past few years, the excitement in scientific, medical, and commercial communities regarding engineered Mabs was on the rise. The list of FDA approved monoclonal antibodies against cancer, viral, and inflammatory diseases is growing rapidly, and more than 30 Mabs are currently at the late stages of clinical trials.

Datamonitor research reports that between 2003 and 2004 the Mab market saw unprecedented growth of 48.1%, exceeding the $10 billion mark. Datamonitor forecasts an annual growth rate of almost 20% for the total Mab market between 2004 and 2010, which far outstrips the pharmaceutical industry as a whole. The number of marketed antibodies is expected to grow more than twice, and 19 new companies are expected to enter the market with Mab products by 2010.

Consistent with reports of recent FDA granted approvals, the engineered Mabs represent over 30% of the biopharmaceuticals in clinical trials. According to Datamonitor, in 2004, the chimeric Mabs dominated Mab market sales at the level of 60.2%. However, the emergence of humanized and fully human Mabs heads a new wave of innovative therapies with rapid market expansion.

References

Wing et al, 1996; Shan et al, 1999; Ammons et al, 2003; Cheng et al, 2003; Hudson and Souriau, 2003; Joosten et al, 2003; Ross et al, 2003; Bang and Keating; 2004; Carter et al, 2004; Gatto, 2004; Kipriyanov and Le Gall, 2004; Sanz et al, 2004; Chatenoud, 2005; Kontermann, 2005.

Key points:

1. The availability of monoclonal antibodies, which target specific cells and individual proteins, is making a significant impact on all fields of medicine, especially oncology, transplantation, and inflammatory and autoimmune diseases.

2. Most of the side effects of the first generation of Mabs, murine Mabs, were successfully overcome by the advent of chimeric, humanized Mabs, and recently developed fully human Mabs.

3. The emergence of the recombinant technologies and new molecular strategies expanded the repertories of Mabs, optimized their molecular architecture and pharmacokinetics, and increased therapeutic efficacy.

Modern innovative methods of Mab production are so sophisticated and multifaceted that they appear to be more an art form than a mere technology. Monoclonal antibodies of almost any required design can be produced that maintain the most crucial feature of antibodies—their specificity to a target antigen. Thus, antibodies are specific and highly powerful tools. However, antibodies represent only one partner of the exclusive "couple" in immunotherapy. The second partner is the antigen itself. In order to ensure the specificity of immunotherapy, the target antigen must be chosen very carefully and wisely, with as much sophistication as the design of specific Mabs itself involves. The crux of the matter is that since the binding of antibody to antigen interferes with cell functioning or even cell death, the antigen must be expressed only on the target cells, or it must be significantly less expressed on cells that are to survive than on the cells targeted for destruction. Otherwise, the therapeutic power of these antibodies may become a dangerous weapon.

Anti-integrin Monoclonal Antibody for Multiple Sclerosis and Crohn's Disease Treatment

The theoretical background for the development of the new agent natalizumab was based on the understanding that the migration of T cells into the brain is the crucial step in the cascade of events involved in the pathophysiology of multiple sclerosis (MS).

Natalizumab prevents immune cells from leaving the bloodstream and reaching the inflamed tissues, and as a result reduces inflammation and demyelination in the MS brain. [Note: Monoclonal antibody against α_4 integrin—natalizumab—has two brand names. Initially it was presented under the name Antegren, which was later changed to Tysabri. The name Tysabri was first used in two clinical trials, AFFIRM and SENTINEL, and the drug received FDA approval under the brand name Tysabri. To avoid confusion and discrepancies in terminology, the generic name "natalizumab," which was never changed, will be used in the following text].

Structure and Mechanisms of Action

Natalizumab is a recombinant Mab against the human α_4 integrin, containing human framework domains and mouse CDR (25%–30% of mouse sequences).

Two monoclonal antibodies against the α_4 integrin have been developed and studied on the animal models and in the clinical trials. Millenium Pharmaceuticals developed Mab, LDPO2, and Elan Pharmaceuticals and Biogen Idec introduced natalizumab. The only conceptual difference between the two compounds consists in the specificity of LDPO2 for blocking the $\alpha_4\beta_7$ integrin interaction with mucosal adhesive cell adhesion molecule-1 (MadCAM-1), whereas natalizumab also inhibits the $\alpha_4\beta_1$ integrin binding with its ligand vascular cell adhesion molecule-1 (VCAM-1).

The blockage of $\alpha_4\beta_7$ integrin interaction with MadCAM-1 by LDPO2 plays the pivotal role in gastrointestinal activated lymphocyte targeting, whereas blocking of $\alpha_4\beta_1$ integrin/VCAM-1 binding by natalizumab appears to be the critical step in reducing the brain inflammation. By binding to $\alpha_4\beta_7$ adhesion molecules in the small intestine and $\alpha_4\beta_1$ molecules in the brain, natalizumab inhibits the pathologies underlying Crohn's disease as well as MS disease. Thus, natalizumab was developed with dual purpose—to treat both the Crohn's disease and MS.

The following presents an analysis of the published clinical trial data as well as some unpublished clinical trial information from companies' press releases.

Clinical Trials of Natalizumab with Crohn's Disease

Four placebo-controlled trials were performed to assess safety and efficacy of natalizumab as a treatment for acute Crohn's disease.

In the first Crohn's disease trial, conducted in the UK, 30 patients received one infusion of natalizumab (n=18) or placebo (n=12). The primary endpoint was a change in the Crohn's disease activity index (CDAI) at week two after a single infusion. There was a statistically significant reduction in the mean CDAI score for the natalizumab-treated group at week two (213 points) compared to the baseline scores (258 points). In the placebo group the differences at week two were not statistically significant (262 points versus 273 points at baseline). At week two, 39% of natalizumab-treated patients versus 8% of the placebo group achieved clinical remission (defined as CDAI scores under 150 points). The beneficial effect of natalizumab was short-lived, and rescue therapy was required for most patients at week four. This first trial showed that one infusion of natalizumab was well tolerated, but data were not robust enough to make a conclusive judgment regarding the therapeutic efficacy of the compound compared to placebo in Crohn's disease patients (Gordon et al, 2001).

The second, more extensive European-Israeli trial included 248 patients who were treated twice at three-week intervals. The trial end-points were defined as the number of patients in clinical remission (characterized by CDAI scores less than 150 points) and the percentage of the patients who achieved 70% or greater drop in CDAI scores at week six. A significantly higher number of patients achieved remission at week six in the natalizumab-treated group compared to the placebo group (44% versus 27%). The reduction in CDAI scores of 70% or greater was observed at week six in all treatment groups with a higher percentage for patients in the natalizumab group compared to the placebo group (71%, 57%, and 38% in the patients receiving 3 mg/kg of natalizumab, 6 mg/kg of natalizumab, and placebo, respectively). This response persisted only through week twelve. The trial demonstrated that two natalizumab infusions were well tolerated and able to mediate an improvement in the patients with moderate-to-severe Crohn's disease. Anti-natalizumab antibodies were detected in 5% of the treated patients (Ghosh et al, 2003).

These two trials provided evidence that one or two natalizumab infusions were well tolerated and effective in short-term treatment of an acute stage of Crohn's disease. Based on the both trials combined data, two long-term trials were designed: ENACT-1 assessing the therapeutic efficacy in the induction of clinical remission and ENACT-2 evaluating the maintenance of the drug-induced remission. Complete results of ENACT-1 and ENACT-2 trials were not published; but in 2005, Van Assche and co-authors described the patient whose participation in both trials resulted in death in 2003. This patient received three monthly 300 mg natalizumab infusions during ENACT-1 trial, followed by placebo injections for nine months during ENACT-2 trial; and then open-label 300 mg natalizimab infusions at a dose 300 mg every four weeks were resumed in response to a relapse of Crohn's disease. In total, the patient has received eight doses of natalizumab (three doses during ENACT-1 trial and five doses during ENACT-2 trial). In **2003,** this patient was diagnosed with fatal astrocytoma, which was not considered as an adverse event directly connected to natalizumab treatment. In **2005,** after retrospective review, this diagnosis was reclassified as JC virus-induced progressive multifocal leukoencephalopathy (PML) associated with natalizumab treatment (Van Assche et al, 2005).

References

Gordon et al, 2001; Van Assche and Rutgeerts, 2002; Dasgupta et al, 2003; Doggrell, 2003; Ghosh et al, 2003; Van Assche et al, 2005.

Key Points:

1. Natalizumab short-term treatment (one or two infusions) was effective and well tolerated in the patients with acute stage of Crohn's disease.

2. There is strong evidence that in the participant of ENACT-1/ ENACT-2 trials lethal progressive multifocal leukoen-cephalopathy was directly associated with natalizumab treatment.

Studies of Natalizumab with Experimental Models of MS

Experimental autoimmune encephalomyelitis (EAE), characterized by inflammation and demyelination within the CNS, is an animal model of MS. Pioneering experiments on EAE were performed in the early 1990s in the framework of collaboration between the laboratories of Dr. Yednock at the Athena Neuroscience and Dr. Steinman at the Stanford University. These studies have shown that anti-α_4 integrin antibody inhibited inflammatory infiltration of the brain and ameliorated paralysis in EAE rats.

Experimental data with blocking interactions between α_4 integrin and VCAM-1 in MS models were encouraging. It has been unequivocally shown that anti-α_4 integrin antibodies injected at the time of EAE induction have arrested the initiation of the clinical manifestations and prevented the development of brain lesions characteristic for the disease. Several studies also reported that anti-α_4 integrin antibodies have inhibited short-term progression of an established EAE and reversed paralysis if given up to a month after the clinical disease onset. At the same time, Brocke and colleagues indicated that anti-α_4 integrin antibodies have had positive therapeutic effects only during the treatment periods. Once Mab administration ended, there was a fast return to the symptomatic EAE. This raised concerns that upon the withdrawal of natalizumab there might be a rebound effect.

On the other hand, it is obvious that the clinically relevant tests for therapeutic potential of the drug in humans are anti-α_4 integrin effects on the MS in progress. For this reason, the ability of the anti-α_4 integrin antibodies to regulate EAE when administered either before or after disease onset was compared during head-to-head experiments performed by Dr. Theien and co-workers. It was shown that the blockage of interactions between α_4 integrin and VCAM-1 prior to the EAE onset inhibited the development and severity of its clinical manifestations. In contrast, administration of anti-α_4 integrin antibodies after the appearance of disease symptoms resulted in the markedly increased relapse rates, an accumulation of CD4+ T cells in the CNS, an augmentation of Th1 responses (peripheral activation of IFN-γ and IL-2 immune responses), and an enhancement of endogenous myelin epitopes spreading. The authors concluded that treatment with anti-α_4 integrin Mab had multiple effects on the immune system, and treatment of ongoing Th-1-mediated autoimmune disease, such as MS, with α_4 integrin antagonists must be used with the caution.

References

Theien et al, 2001; Steinman, 2005.

Key points:

1. In MS animal models, the blockage of α_4 integrin/VCAM-1 interactions by antibodies against α_4 integrin prevents the development of EAE.

2. Antibodies against α_4 integrin are much more effective for preventing EAE than for restraining clinically established disease.

3. For some conditions, the treatment of ongoing EAE with anti-α_4 integrin antibodies increases the disease activity.

4. Anti-α_4 integrin antibodies reduce progression of previously established EAE only for the duration of the treatment periods.

Clinical Trials of Natalizumab with MS

Several randomized, double-blind, placebo-controlled trials were conducted to evaluate the therapeutic efficacy and tolerability of natalizumab in patients with MS.

Phase I Trial
(Sheremata et al, 1999; Minagar et al, 2000).

Trial Design

Phase 1, randomized, placebo-controlled study of a single intravenous natalizumab dose (between 0.01 and 3.0 mg/kg of body weight) was performed with 28 stable RRMS and SPMS patients. For extended follow-up, the patients were assigned to one of three groups: group I (7 patients) received placebo; group II (9 patients) received a low dose (0.01–0.3 mg/kg) of natalizumab, and group III (12 patients) received a high dose (1.0–3.0 mg/kg) of natalizumab. The patients were followed prospectively for 8 weeks, and subsequent data collection was performed retrospectively.

NOTE: The definition of relapses that was chosen in this trial, "An exacerbation was defined as appearance of a new symptom or reappearance of an old MS symptom of central nervous disease attributable to MS, with confirmation of new neurological signs by examination resulting in an increase on the EDSS by 1 grade," is different from the definition of repalses suggested by Schumacher and co-authors, which is used in the most MS trials.

Rate of Clinical Relapses

Investigators reported an apparent reduction in the rates of annual relapses in RRMS and SPMS patients after a single infusion of natalizumab. In the placebo group, relapses occurred in 5 out of 7 patients (71%). In group II, relapses occurred in 2 out of 9 patients (22%), and in group III, only 1 out of 12 patients (8%) who received 3 mg/kg of natalizumab, experienced a relapse after 5 months.

The investigators concluded that the blocking $\alpha_4\beta_1$ integrin by a single infusion of anti-α_4 integrin Mab had the dose dependent (from 0.01 to 3.0 mg/kg of body weight), extremely strong, and long-lasting (one year) beneficial effects in the patients with MS.

Key Points:

1. The small number of patients in the study warrants caution in the interpretation of these results.

2. The observation of the long-lasting reduced exacerbation rates in MS patients following a single dose of natalizumab contradicted to the other study described below. Authors of this study suggested, "Differences in definition of relapses in the two studies might explain these discrepancies."

UK Trial
(Tubridy et al, 1999).

Trial Design

UK study was conducted in eight clinical centers in the United Kingdom. In total, 72 patients with active MS were assigned to receive either natalizumab or placebo. The natalizumab group included 25 patients with relapsing remitting MS (RRMS) and 12 patinets with secondary progressive MS (SPMS). The control group included 28 RRMS patients and 7 SPMS patients. Intravenous infusions of 3 mg/kg of body weight of natalizumab or placebo were administered on two occasions four weeks apart. The cumulative number of new active and new enhancing lesions on MRI, the number of clinical relapses, and changes in disability status during weeks 1 to 24 from the start of treatment were chosen as the treatment efficacy parameters.

Number of Lesions. 1–24 weeks (1–12 and 12–24 weeks)

After the first 12 weeks of treatment with natalizumab, the number of new active lesions per patient identified by MRI was significantly reduced: 1.8 for the natalizumab group versus 3.6 for the placebo group (p=0.042). The number of new enhancing lesions was also reduced: 1.6 for the natalizumab group compared to 3.3 for the control group (p=0.017).

From weeks 12 to 24 after the first dose, there were no significant differences in the mean cumulative number of new enhancing lesions between the natalizumab and the placebo groups of patients. The mean number of enhancing lesions existing at the baseline was similar in both treatment and the control groups during the entire trial period.

Rate of Clinical Relapses. 1–24 weeks (1–12 and 12–24 weeks)

After the first 12 weeks of treatment, no significant differences were found between the treatment and the control groups in the number of patients with at least one MS relapse.

During the second 12 weeks after the administration of the first dose of natalizumab, the percentage of patients with acute MS relapses was significantly higher (38%) in the natalizumab treatment group compared to (9%) in the placebo group (p=0.005). Moreover, during weeks 12 to 24 after the first dose, 21% of patients treated with natalizumab required steroid therapy, whereas none of the patients in the placebo group did. In the natalizumab group, 11% of patients were hospitalized for exacerbated symptoms. No patients in the placebo group were hospitalized during the course of the study.

There were no changes in EDSS scores in either group during the 24-week study period, but, actually, it was unrealistic to expect them for such a short-term trial.

Adverse Events

There were no significant differences in the incidence of general adverse events between two groups except for fatigue, which was higher in the natalizumab group, particularly during the second 12 weeks. Between weeks 1 and 12, the natalizumab-treated group experienced significant leukocytosis (number of leukocytes was 56% to 60% greater than that in the placebo group). Blood counts returned to the baseline levels between 20 and 24 weeks. Antibodies to natalizumab were detected in 11% of natalizumab-treated patients; antibody titers ranged from 1.3 μg/ml to 5 μg/ml. Severe adverse events were reported in four natalizumab patients, and in none among the patients receiving placebo.

Key points:

1. The significant reduction in the number of new active lesions and new enhancing lesions over the first 12 weeks of the study suggested that the new lesion formation may be suppressed by two infusions of natalizumab. This effect is maintained for a short period of time after the treatment cessation.

2. The unchanged number of enhancing lesions persistent from the baseline in both the natalizimab and the placebo groups over the entire trial period indicated that natalizumab has no apparent effect on the lesions which were already established prior to the treatment. These data are concordant with the results obtained by Dr. Theien and co-workers, who had previously suggested that antibodies against α_4 integrin

are much more effective in the prevention of EAE compared to their therapeutic efficacy on an ongoing process.

3. The significant increase in the relapse rate, the necessity for steroid treatment, and the increased incidence of hospitalizations in the natalizumab cohort compared to the placebo group during the period between 12 and 24 weeks after the discontinuation of the treatment corresponded to data presented by Dr. Thien et al. about an increased exacerbation rates in the EAE model after the cessation of antibody infusions. These results raise serious concerns regarding dramatically increased rebound effects.

4. Overall, the positive effects of short-term natalizumab treatment were modest and transient.

5. If there were no alerting data about the clinical decline in MS patients after the treatment cessation, the reduction in the number of new active lesions and new enhancing lesions over the first 12 weeks after the first dose, at the best may be considered supportive for short-term natalizumab administration at the acute stages of MS. However, given the rebound effects upon natalizumab discontinuation, even this indication appears imprudent.

Despite its several serious shortcomings and apparently disturbing outcome, the drug developers considered the results of the phase II UK trial to be promising for the reduction of the disease activity. Therefore, the phase III studies were initiated to evaluate natalizumab as a new treatment option for the patients with MS.

The International Natalizumab Multiple Sclerosis Trial (INMST)

(Miller et al, 2003).

Trial Design

The INMST was conducted in twenty six clinical centers in the US, Canada, and the United Kingdom. A total of 213 patients were randomized to one of three treatment regimens: a cohort of 68 patients—47 RRMS patients and 21 SPMS patients—were treated with the dose of 3 mg of natalizumab per kg of body weight; 74 patients—52 RRMS patients and 22 SPMS patients—received 6mg of natalizumab per kg of body weight; and a cohort of 71 patients—45 RRMS patients and 26 SPMS patients—were given placebo. Each patient received intravenous infusion of natalizumab or placebo every four weeks for six months. After the treatment cessation, the patients were monitored for an additional six months to determine the efficacy of treatment and occurrence of adverse events. The trial end-points included the number of brain lesions on monthly MRI during six months of the treatment and during six month of the follow-up period and clinical outcome measures: subjective relapses reported by patient, objective relapses evaluated by neurologists, self-reported well-being, and changes on EDSS.

Number of New Enhancing Lesions during Six Months of Treatment and Six Months of Follow-up Periods

The study demonstrated that six months of natalizumab treatment resulted in a statistically significant reduction in the number of new enhancing lesions in a mixed RRMS and SPMS population compared to a mixed RRMS and SPMS population in the placebo group. To the end of treatment period, the mean number of new enhancing lesions per patient was 0.7, 1.1, and 9.6 in natalizumab

3 mg/kg, natalizumab 6 mg/kg, and the placebo groups, respectively (p=0.001).

The effects of natalizumab treatment were even more striking in the subgroup of RRMS patients: the mean number of 12.1 new enhancing lesions per patient in the placebo group compared to 0.6 lesions in the 3 mg/kg and the 6 mg/kg natalizumab groups (p<0.001). Although the results in the SPMS subgroup of patients were less impressive, there were still statistically significant differences between the 3 mg/kg natalizumab treated group and the placebo group (1.0 lesion in the group given 3 mg of natalizimab as compared with 5.4 lesions in the placebo group; p=0.005). A positive trend was observed in the group given 6 mg/kg of natalizumab, which had the mean number of 2.0 new enhancing lesions compared to 5.4 lesions per patient in the placebo group (p = 0.08).

During the six-month follow-up period, the number of new enhancing lesions was similar in all three groups: 2.5, 2.3, and 2.4 lesions per patient in the natalizumab 3 mg/kg, 6 mg/kg, and the placebo groups, respectively.

Scans Showing Activity (%) during Six Months of Treatment and Six Months of Follow-up Periods

During the six-month treatment period, the percentage of scans showing activity in the natalizumab-treated groups was statistically lower than that in the placebo group: 9%, 11%, and 39% in the natalizumab 3 mg/kg, natalizumab 6 mg/kg, and the placebo groups, respectively. After six months of follow-up, the percentage of scans with new enhancing lesions was approximately the same in all three groups: 41%, 36%, and 40% in the natalizumab 3 mg/kg, natalizumab 6 mg/kg, and the placebo groups, respectively.

Analysis of the scans with different numbers of new enhancing lesions by the end of six months of treatment for the natalizumab and the control groups has shown an unusual distribution. The percentage of the scans with 1–3 lesions was approximately equal for all three groups: 25%, 21%, and 27% in the natalizumab 3 mg/kg, natalizumab 6 mg/kg, and the placebo groups, respectively. There were no scans with 7–9 new enhancing lesions in any of the three groups. The scans with 10–12 new enhancing lesions were reported in 1% of the natalizumab 3 mg/kg group, none of the natalizumab 6 mg/kg treated patients, and in 4% of the patients receiving placebo. The substantial differences were found only in the percentage of the scans with 4–6 and with more than 12 new enhancing lesions. The percentage of the scans with 4–6 new enhancing lesions was 1%, 7%, and 18% in the natalizumab 3 mg/kg, natalizumab 6 mg/kg, and the placebo groups, respectively. The percentage of scans with more than 12 new lesions was 1% for both natalizumab 3 mg/kg, and 6 mg/kg groups, and 20% for the placebo group.

Rate of Clinical Relapses during Six Months of Treatment and Six Months of Follow-up Periods

During the six-month treatment period, the total number of relapses and relapses per patient were lower in the natalizumab-treated groups compared to the placebo group: 0.26, 0.20, and 0.51 relapses per patient in the natalizumab 3 mg/kg, natalizumab 6 mg/kg, and the placebo groups, respectively. During the treatment period, the number of patients with relapses in the natalizumab-treated groups was significantly lower than in the placebo group. In both the natalizumab 3 mg/kg and 6 mg/kg groups, only 19% of patients had relapses, whereas 38% of those assigned to the placebo group have relapsed (p=0.02).

Over the six-month follow-up period, the number of relapses per patient was similar in all three groups: 0.36 relapses per patient in the natalizumab 3 mg/kg group, 0.39 in the natalizumab 6 mg/kg group, and 0.35 relapses in the placebo group. During the follow-up period, no significant differences in the percentage of the patients experiencing acute relapses was found between the three groups: 31% in the natalizumab 3 mg/kg, 34% in the 6 mg/kg group, and 35% in the placebo patients.

The 100 mm visual-analogue scale was used to assess self-reported well-being. The patients marked points along the 100 mm line reflecting their own evaluation of overall well-being at baseline and after six months of treatment. The placebo group reported a slight worsening in well-being with a mean decrease by 1.38 mm, whereas the natalizumab groups indicated an improvement with a mean increase of 9.49 mm in the group given 3 mg/kg and 6.21 mm in the group given 6 mg/kg of natalizumab. Although the authors used this test for characterization of the frequency of relapses, the severity of fatigue and clinical stability, the perception of well-being should be interpreted with caution, since the visual-analogue scale assessment of well-being is a subjective parameter, which has not been widely used in clinical practice. Assessment of well-being for the follow-up period was not published.

No significant changes in EDSS scores were observed in any group during six months of treatment.

Adverse Events

There were no significant differences in the incidence of adverse events between three groups except for a trend toward an increased rate of infections in the natalizumab—treated groups. Total white cell counts increased in both natalizumab groups during the first month of treatment and remained elevated by the first two months of the follow-up period. However, the mean values were not outside of the normal range. According to Doggrell, and Miller with co-authors, antibodies against natalizumab developed in 11% of the patients during the treatment and post-treatment periods.

Key Points:

1. The INMST study demonstrated that six months of natalizumab treatment resulted in statistically significant reduction in the number of new enhancing lesions, the percentage of scans showing activity, the number of relapses, and the percentage of the patients with clinical relapses.

2. All beneficial effects persisted only during the natalizumab injections on a regular basis.

3. After treatment cessation, there were no significant differences among the natalizumab-treated and placebo groups for all MRI and clinical parameters.

Message to the Readers

The findings in this study may have been adversely influenced by the unusual patterns of disease indexes reported in the placebo group.

After six months of placebo infusions, the mean number of new enhancing lesions was 9.6 per patient, which correlated well with the high rate of disease activity in the patients enrolled in the study (average formation of 1.6 new lesions per patient each month). Therefore, 9.6 lesions for six months without treatment would be expected. However, if the high rate of new lesions during the placebo infusions could be explained by high activity of the disease process, the dramatic decrease in the number of new enhancing lesions (2.4) six months after the placebo infusions were stopped would be inexplicable. All other parameters measured in the placebo group showed the same trend—relatively poor during placebo injections and radically improved after it was discontinued. The mechanisms behind the dramatic decrease of MS activity in the placebo group after placebo infusions ended remain unclear. It is unlikely that the initial infusion with a placebo solution and its cessation could have such remarkably positive effects on the course of MS. Given the striking fluctuations in the placebo group, it is difficult to interpret the overall findings during the treatment and post-treatment periods.

The second aspect, which requires clarification, pertains to the unusual grouping of the scans with different number of new enhancing lesions. There were approximately equal percentage of the patients in all three study arms with 1–3, 7–9, and 10–12 new enhancing lesions. Significant differences between the treatment and the placebo groups in the percentages of active scans was reported only for the patients with 4–6 and over 12 new enhancing lesions. It is difficult to explain the phenomenon of such a sharp bimodal distribution of the new enhancing lesion formation.

AFFIRM Monotherapy Trial

(Data are not published; in the press release, the companies Elan Pharmaceuticals and Biogen Idec presented information regarding the one-year MRI measurements and clinical outcomes)

Trial Design

AFFIRM trial was a two-year study conducted in 99 sites worldwide; 942 MS patients were randomized to receive either a fixed 300 mg intravenous dose of natalizumab (n=627) or placebo (n=315) every four weeks. The study enrolled patients who had neither received any IFN-β nor glatiramer acetate for at least the previous six months. Approximately 94% of the trial participants had never been treated with either of these agents. Neurological evaluations were performed every 12 weeks and at the time of suspected relapse. MRI evaluations for T1-weighted gadolinium (Gd)-enhancing lesions and new or newly enlarging T2-hyperintense lesions were performed annually.

Number of Lesions

After one year of treatment, MRI scans detected no T1-weighted Gd-enhancing lesions whatsoever for 96% of natalizumab-treated patients compared to 68% in the placebo—treated group (p<0.001). One Gd-enhancing lesion was detected in 3% of natalizumab-treated patients compared to 13% in the placebo group. Two or more Gd-enhancing lesions were detected in 1% of natalizumab-treated patients compared to 19% in the placebo group.

In the group treated with natalizumab, 60% of the patients developed no new or newly enlarging T2-hyperintense detectable lesions compared to 22% of the placebo-treated patients (p<0.001). The percentage of the patients with one T2-hyperintense lesion was fairly comparable in both groups: 18% in the natalizumab group and 13% in the placebo group. The number of patients with two T2-hyperintense lesions on imaging was also approximately equal: 6% in the natalizumab group and 7% in the placebo group. MRI found three or more T2-hyperintense lesions in 16% of the natalizumab group, which was significantly lower than the reported 58% of the patients from the placebo group.

Rate of Clinical Relapses

An annualized relapse rate of 0.25 was reported for the natalizumab-treated patients versus 0.74 for the placebo-treated patients, implying that during one-year treatment natalizumab reduced the rate of clinical relapses by 66% relative to placebo (p<0.001). The proportion of patients who remained relapse free was 76% in the natalizumab group compared to 53% of the control group (p<0.001).

SENTINEL Add-on Trial

(Data are not published; in the press release, the companies Elan Pharmaceuticals and Biogen Idec presented information regarding the one-year MRI measurements and clinical outcomes)

Trial Design

SENTINEL trial was designed to be a two-year study conducted in 123 sites worldwide; 1,171 Avonex-refractory MS patients were randomized to receive either 300 mg of natalizumab (n=589) or placebo (n=582) every four weeks for up to 28 months. All patients continued to receive Avonex in standard regimen. Neurological status was assessed at the previously specified time. MRI evaluations for T1-weighted Gd-enhancing lesions and new and newly enlarging T2-hyperintense lesions were performed annually.

Number of Lesions

By the end of the first year of treatment, 96% of the Avonex/natalizumab-treated patients had no MRI Gd-enhancing lesions compared to 76% of the Avonex/placebo-treated group (p<0.001). One Gd-enhancing lesion was detected in 3% of the natalizumab-treated patients compared to 12% in the placebo group. Imaging detected two or more Gd—enhancing lesions in 1% of the Avonex/natalizumab patients versus 12% in the Avonex/placebo group.

In the Avonex/natalizumab group, 67% of patients developed no new or newly enlarging T2-hyperintense lesions compared to 40% of the

Avonex/placebo patients (p<0.001). The percentage of the patients with one T2-hyperintense lesion was comparable in both groups: 26% in the Avonex/natalizumab group and 29% in the Avonex/placebo group. The percentage of the patients with two new lesions was also similar for both groups: 4% in the Avonex/natalizumab cohort and 10% in those treated with Avonex plus placebo. The percentage of the patients with three or more T2-hyperintense lesions was significantly lower for the Avonex/natalizumab-treated patients compared to the Avonex/placebo patients (3% versus 21%).

Rate of Clinical Relapses

The addition of natalizumab to the Avonex regimen resulted in 54% fewer clinical relapses than treatment with Avonex alone (p<0.001). The annualized relapse rate was 0.36 for patients receiving natalizumab and Avonex, whereas subjects treated with Avonex plus placebo had 0.78 relapse rate. The proportion of patients who remained relapse free was 67% in the Avonex/natalizumab group compared to 46% in the Avonex/placebo controls (p<0.001).

Adverse Events

Common adverse events associated with natalizumab treatment included the urinary and lower respiratory tract infections. Serious infections occurred in 2.1% of the natalizumab-treated patients compared to 1.3% of the placebo-treated patients. The bacterial infections such as pneumonia and urinary tract infections responded appropriately to antibiotics. Antibodies against natalizumab were detected in approximately 10% of the patients.

Serious Adverse Events

On February 2005, two years after the commencement of the SENTINEL trial, two cases of progressive multifocal leukoencephalopathy, one of which was fatal, were reported in patients from the Avonex plus natalizumab-treated group.

Key Points:

1. The results of both trials were strikingly similar. Statistically significant percentage of patients in the natalizumab-treated groups had no new or newly enlarging lesions compared to the placebo groups. Thus, during one year of the treatment, in many cases, natalizumab was able to successfully inhibit the formation of new lesions.

2. The smaller percentage of the patients with three or more T2-hyperintense lesions in the natalizumab-treated groups compared to the placebo groups demonstrated that during one year of treatment natalizumab restrained the active inflammation process in MS patients.

3. The percentage of the natalizumab-treated patients with zero, three, or more T2-hyperintense lesions was lower compared to the placebo-treated patients. In contrast, in both trials the percentage of the patients with one or two T2-hyperintense lesions was comparable for the natalizumab and placebo groups. This fact raises the question about the possibility of varying sensitivities to the natalizumab treatment at different stages of the disease activity.

4. During one year of the treatment, natalizumab controlled the development of clinical relapses with high efficacy.

5. Two patients who participated in SENTINEL trial were affected by progressive multifocal leukoencephalopathy. For these two cases, there is strong evidence of direct association between the development of progressive multifocal leukoencephalopathy and natalizumab treatment.

Accelerated FDA Approval of Natalizumab (Tysabri) and Withdrawal of Natalizumab (Tysabri) from the Market

Elan Pharmaceuticals and Biogen Idec filed a Biologics License Application with the Food and Drug Administration based on the unpublished, one-year interim analyses of two ongoing Phase III two-year clinical trials, AFFIRM and SENTINEL.

On November 23, 2004, FDA approved Tysabri (natalizumab), formerly referred to as Antegren, as treatment for the relapsing form of MS. The FDA granted accelerated approval for Tysabri following a priority review based on the one-year unpublished data from the two Phase III two-year studies, the AFFIRM monotherapy trial and the SENTINEL add-on trial with Avonex plus Tysabri.

Before accelerated FDA approval of the drug was granted, the affairs about natalizumab were widely advertised. The drug manufacturers raised high hopes and issued assurances gaining the attention of the public, medical, and business communities. The promises were made that natalizumab would begin a new era of treatment offering MS patients genuine improvements over currently available drugs. Market analysts were convinced that natalizumab would quickly attain "blockbuster status, dominate the market, and become the biggest MS drug of all times."

Tysabri was withdrawn from the market on February 28, 2005, following reports of the two confirmed cases of progressive multifocal leukoencephalopathy in the SENTINEL clinical trial.

References:

Sheremata et a, 1999; Tubridy et al, 1999; Minagar et al, 2000; Theien et al, 2001; Doggrell, 2003; Miller et al, 2003; Sheridan, 2004; Biogen Idec, Elan press releases, 2004 and 2005; Carter et al, 2004; Drazen, 2005; Kleinschmidt-Demasters and Tyler, 2005; Langer-Gould et al, 2005; Sheridan, 2005; Steinman, 2005.

Message to the Readers

The lessons gained from the dramatic withdrawal of Tysabri from the market should be hearkening to other companies, which are developing drugs of the same class. Further concerns should be raised regarding their safety. The mechanisms of action underlying α_4 integrin antagonists are currently the focus of intense scrutiny (Sheridan, 2005).

Natalizumab developers emphasized only one aspect of the natalizumab action: the prevention of the migration of circulating immune cells into the CNS. This aspect is certainly legitimate and beneficial for the treatment of MS. Unfortunately, the anti-α_4 intergin antibodies are not specific only for the lymphocytes activated in the course of MS. As discussed previously, α_4 intergins are constitutively expressed on the surface of multiple cells and participate in a broad range of vitally important processes. Because treatment of MS necessitates the prolonged or even life-long treatment, extended interruption of the various α_4 intergin-mediated functions, theoretically, may be implicated in several pathologies, including:

- Greatly increased susceptibility to the bacterial and viral infections
- Defective intergin-dependent regulation of hematopoietic processes
- Increased risk of hematological and other malignancies
- Alterations in spermatogenesis, fertility, and embryonic development

At the same time, the clinical efficacy of natalizumab is impressive. During treatment periods, there were striking reduction in the rates of clinical relapses and brain lesion formation. According to the press release by Elan Pharmaceuticals and Biogen Idec, natalizumab treatment decreased the risk of disability progression by 42% compared to the placebo. It is possible that repeated detailed retrospective analyses of all trials (no matter how small they were) may help to identify the population of patients whose benefits from natalizumab treatment would outweigh the risks. Careful examination of the multiple effects of α_4 intergin antagonists on the vitally important systems is mandatory in order to assess the extent of the natalizumab benefit-risk ratio.

Analysis of progressive multifocal leukoencephalopathy and its possible association with natalizumab treatment is presented in the following chapter.

Chapter 4

Progressive Multifocal Leukoencephalopathy

Definition

Progressive multifocal leukoencephalopathy (PML) is an opportunistic viral demyelinating disease of the CNS, which occurs almost exclusively in the setting of severe or prolonged immunosuppression.

The JC virus (JCV) is the known etiologic agent of PML. JCV selectively targets the glial cells: it infiltrates oligodendrocytes and, to a lesser degree, infects astrocytes. PML results in direct lytic destruction of the glial cells followed by the secondary demyelination and progression to the multifocal lesions.

The immunosuppression provides ideal conditions for the JCV replication and proliferation in the CNS.

References

Greenlee, 1998; Baum et al, 2003; Du Pasquier et al, 2003; Eash et, 2004; Seth et al, 2004.

Symptoms

The only known function of oligodendrocytes is the production of myelin. Astrocytes used to be thought of as a mere support for the neuron's functioning. Over the past several years, however, the experiments performed by Blomstrand with co-workers, Pfriege and Barres, and Rothstein and co-authors indicated that astrocytes execute much more independent functions: partially metabolize glucose in the CNS; contain a high affinity glutamate transporters, which are critical in the maintaining the extracellular glutamate concentration at sub-excitotoxic levels and prevention of the cell death; and integrate neuronal inputs by modulating the composition, volume, and the concentration of ions and neurotransmitters. Thus, the destruction of oligodendrocytes with secondary demyelination leads to a delay of signal transmission from the CNS to the target cells, whereas the destruction of astrocytes impairs the integration of several brain functions.

Once the JCV enters the brain, astrocytes and oligodendtocytes support the multiplication of virus. Therefore, although PML is defined as a demyelinating disease, JCV infection of both types of the neuroglial cells—oligodendtocytes and astrocytes—impairs other neuroglial functions in addition to the production and maintenance of myelin.

JCV damage to both types of the microglial cells manifests with a variety of PML symptoms. The clinical picture of PML is dominated by an insidious onset and rapidly progressive symptoms. Neurological deficits may include focal or diffuse neurological signs and mental impairment. PML is commonly characterized by the generalized motor weakness, weakness in the arms or legs (usually unilateral), and hemiparesis (which may not be present at the early stages but eventually occurs in 75% of JCV cases at the later stages). Ataxia and vestibular dysfunction are common. Visual disturbances are present in 45% of cases typically as homonymous hemianopia, blurred or double vision, or complete loss of vision in one eye. Furthermore, 38% of the patients have a variety of symptoms related to a decline in mental functioning—confusion, disorientation, personality changes, and dementia.

JC virus may spread to the cerebral hemispheres, brain stem, cerebellum, and cervical spinal cord. PML symptoms vary depending on the location of the brain lesions and the rate of virus dissemination. Rate of virus dissemination may correspond to the regional variability of oligodendrocyte densities, for instance, in the heavily myelinated human pontocerebellar tract the oligodendrocyte count is estimated to be $64,000/mm^3$, so virus infection in this area may be more pronounced. Focal signs tend to be related to the posterior brain regions (e.g. occipital lobes) leading to the presenting symptoms of hemiparesis, ataxia, aphasia, and cortical blindness. Early brainstem damage may be presented by occulomotor palsies, stupor, clouding of the consciousness, central vestibular syndrome, and dysphagia. The symptoms associated with astrocyte damage are often characteristic for the stratum radiatum impairment in the hippocampus CA1 area because astrocytes are the predominant glial subtype there filling between 4% and 8% of this brain region volume.

References

Major et al, 1992; von Einsiedel et al, 1993; Rothstein et al, 1996; Pfriege and Barres, 1997; Blomstrand et al. 1999; Fazakerley and Walker, 2003; Gasnault et al, 2003; Hansson and Rönnbäck, 2003; Roos, 2005.

PML Sub-types

Atypical and severe forms of PML are distinguished based on the pathomorphology and symptoms of the disease. One of the atypical forms of the disease, though containing all of the key elements of the classic PML pathology, is characterized by unifocal brain lesion. The leading pathological feature of this severe type of PML consists of particularly intense brain tissue destruction and extremely extensive necrotizing cavitation. The other distinguishing feature of severe PML is the strong topographic prevalence of the pathological processes extending to either the brain hemispheres or to the cerebellum.

In about 13% of PML patients (HIV-negative, HIV-positive, and AIDS patients) the disease is associated with the transient increase of immune activity and an enhancement of the brain inflammation. This inflammatory form of PML generally has more favorable prognosis. It was shown that an immune restoration following effective treatment of AIDS has been accompanied by a significant decrease in infectious morbidity and in prolonged survival in AIDS/PML patients. However, despite the documented benefits, in some patients with AIDS-associated PML who were treated with highly active antiretroviral therapy (HAART), the rapid recovery of immune system functions paradoxically exacerbates the clinical course of PML, sometimes very severely. This exacerbation of PML is thought to be associated with a rapid immune reconstitution, particularly in the patients with very low baseline CD4+ T cell counts who promptly responded to the anti-HIV therapy in terms of increased CD4+ T cell numbers. It was suggested that a gradual reversal of the immune deficiency would possibly produce a better outcome. The inflammatory PML is considered as a separate clinical entity among the expanding range of the diseases commonly referred to as "immune reconstitution syndrome" or "immune restoration disease." The inflammatory PML occurs due to an exaggerated immune response or dysregulated restitution of the immune system.

SENTINEL trial investigators recently reported the patient's case with initially clearly demonstrative clinical course of the classic PML, which was later transformed into the inflammatory form of PML. The MS patient, a participant of the SENTINEL trial, received 28 infusions of natalizumab and during treatment was affected with typical rapidly progressive PML. Three months after cessation of natalizumab treatment, the immune reconstitution syndrome developed, which was characterized by a widespread inflammation in the CNS, increased MRI lesions, and a dramatic clinical deterioration. Treatment with systemic cytarabine therapy improved the patient's condition. The exact causes for this improvement are not clear. It is possible that cytarabine sufficiently diffused into the brain of this patient due to increased permeability of the blood-brain barrier (BBB), which is common for multiple sclerosis (usually penetration of cytorabine into the central nervous system is negligable); cytorabine may have directly killed JCV inside of the brain, or cytorabine may have reduced the central nervous system inflammation because of its strong myelosuppressive properties. It is also possible that the patient's recovered immune system itself participated in the elimination of virus. This patient is one of three patients in whom rapidly progressive PML occurred during natalizumab clinical trials, and is the one who survived. The case confirmed previous suggestions that the inflammatory PML has a more favorable prognosis than classic PML. However, patients can die during the course of the reconstitution syndrome, as this patient almost did (Langer-Gould et al, 2005).

References

Schmidbauer et al, 1990; Mossakowski and Zelman, 2000; Cinque et al, 2001; Miralles et al, 2001; Du Pasquier et al, 2003 ; Du Pasquier and Koralik, 2003 ; Gray et al, 2003; Hoffmann et al, 2003. Langer-Gould et al, 2005.

Diagnosis

Dr. Cinque and co-authors proposed that, according to the diagnostic criteria, PML cases may be categorized as follows: "histology confirmed" PML with evidence of JCV infection in the brain; "laboratory confirmed" based on the detection of JCV DNA in the cerebrospinal fluid (CSF); and "possible" PML based on the presence of typical clinical and radiological picture, but without a measurable detection of JCV. The disease activity is determined by the indicators including neurological score, MRI findings, and virological marker (JCV DNA in CSF). These parameters have to be assessed, at least, every 3 months until the disease arrest or death of the patient.

The radiologic neuropathology is characterized by both small foci and large areas of demyelination in the cerebrum, brain stem, and cerebellum. MRI reveals multifocal asymmetric white matter lesions. Computed tomography (CT) scans, which are less sensitive than MRI for PML diagnostics, often show hypodense nonenhancing white matter lesions.

Molecular analysis of JCV DNA in the CSF by the polymerase chain reaction (PCR) is recognized as a sensitive and specific method for human polyomaviruses detection. JCV DNA was identified in the majority of CSF samples from the AIDS-associated PML patients, but not in the CSF from non-PML

AIDS cases. However, the PCR method is not 100% diagnostic for PML because some PML patients are JCV-negative. JCV DNA detection in the peripheral blood mononuclear cells (PBMCs) and in urine did not differ between AIDS patients with or without PML and healthy individuals.

Brain biopsy remains the most precise method to validate PML diagnosis. Postmortem histological examination is the definitive confirmatory diagnostic approach for the epidemiological studies.

References

Ferrante et al, 2001; Whiley et al, 2001; Cinque et al, 2003; Fazakerley and Walker, 2003.

Epidemiology

Rates of PML Incidence in General Population, 1958–1984

PML was first described in 1958 by Astrôm, Mancall, and Richardson. Despite the fact that the seroprevalence of latent infection with JCV exceeded 80% of the human population, until the AIDS pandemic PML remained a virtual medical curiosity. In large review of data from 1958 through 1984, Brooks and Walker found as few as 230 published and unpublished pathologically confirmed cases of PML. Approximately 95% of these 230 PML patients had an underlying condition generally associated with the immunosuppression of which only 3% were AIDS-related.

Rates of PML Incidence in Immunocompromised Patients, 1984-Present Time

AIDS

AIDS was first recognized in the USA in the summer of 1981. In 1983, HIV was isolated, and by 1984, this virus was clearly demonstrated to be the causative agent for AIDS. The evolving AIDS pandemic quickly changed the epidemiology of PML. For example, almost 20-fold increase in the incidence of PML in south Florida was observed during the four-year period of 1990–1994 compared to 1980–1984. During the four-year period of 1990–1994 a staggering 98.7% of PML cases were AIDS-related compared to just 3% of PML attributed to AIDS by the 1958–1984 data. Currently, in the USA and in other developed countries, AIDS is the underlying condition in 85%–90% of PML cases; and PML ultimately occurs in 3–10% of all AIDS patients during the course of their illness. Unlike the incidence of many opportunistic infections associated with AIDS (cerebral toxoplasmosis, CMV encephalitis, and HIV encephalitis), which decline following the introduction of HAART for AIDS treatment, the incidence of PML does not significantly differ between the pre-HAART and the HAART periods. Only one research (Gray and co-workers) reported decreased incidence of PML between the years 2000 and 2002.

References

Holman et al, 1998; Antinori et al, 2001; Berger, 2003; Ferrante et al, 2003; Gray et al, 2003; Eash et al, 2004.

Immunosuppressive Treatment

As mentioned previously, in 1990–1994, 98.7% of all PML cases were AIDS-related. Currently, the proportion of AIDS-related PML is reduced to 85–90%. The difference is made up by the appearance of other PML predisposing factors—such as new therapies for cancer and immune diseases, which lead to the immunosuppression and increase the risk of susceptibility to PML. Two pertinent examples are presented below.

Rituximab is monoclonal antibody against CD20 antigen, which is expressed on the surface of malignant cells as well as on all normal B lymphocytes. Rituximab is indicated for the treatment of patients with CD20+ B cell non-Hodgkin's lymphoma. Although rituximab causes depletion of normal circulating and tissue-based B cells followed by a significant reduction of serum antibodies (IgG and IgM), lymphoma treatment with rituximab alone does not results in a dramatic increase of infections. B cells recover to the pre-treatment values within 12 months of therapy cessation. At the same time, combination of rituximab and autografting resulted in more profound immunosuppression and the development of viral diseases including PML. It was shown that during time interval between 5 and 20 months after undergoing autologous peripheral blood stem cell transplantation in combination with rituximab, 4 out of 62 patients (6.4%) were reported to develop severe viral infections (two cases of PML and two cases of CMV disease) compared with no cases of PML or CMV among the 276 patients not receiving immunosuppressive therapy (p<0.001). The second example, which was discussed in the previous chapter, concerned the development of three PML cases associated with the natalizumab treatment. Two cases of PML occurred among 589 MS patients (0.34%) after two years of treatment with natalizumab (Tysabri) in combination with Avonex, and one case of PML developed in patient with Crohn's disease who received natalizumab alone.

References

Goldberg et al, 2002; Boye et al, 2003; Biogen Idec, Elan Pharmaceutic Press release, 2005; Kleinschmidt-De-Masters and Tyler, 2005; Langer-Gould et al, 2005; Van Assche et al, 2005.

Rates of Death. Natural History versus HAART-treated Patients with AIDS

PML is a fatal disease. As PML natural history shows, the disease progresses very aggressively, and median survival period is on average two-six months. To date, there is no etiological or pathophysiological treatment for PML.

Analysis carried out in the USA from 1979 through 1994 showed that during this period the PML death rate increased by more than 16 times: from less than 0.2 per million persons before 1984 to 3.3 per million persons in 1994. As mentioned earlier, the increase was mainly attributable to HIV infection. The death certificates from 1991 through 1994 documented that 89% of PML patients were HIV positive. PML is emerging as a persistent cause of the death in AIDS patients and has the lowest one-year survival probability of any other cerebral disorder.

The introduction of HAART for AIDS treatment resulted in a dramatic improvement of AIDS clinical course. Although AIDS-associated PML remains the life threatening disease with poor prognosis, nevertheless, PML patients with AIDS benefit from HAART in terms of a prolonged survival. Several studies worldwide have shown that the median survival time for HAART-treated PML/AIDS patients was significantly prolonged (up to 2 years), and the risk of death within six months after diagnosis of PML was reduced (for 50% of these patients). Thus, it was proved that HAART, on average, prolonged survival for PML patients.

However, approximately half of the patients did not benefit from this treatment. Some PML patients became long-term survivors, whereas others had a rapid fatal outcome.

Several determinants influence the survival of PML patients and may be considered as prognostic factors for the course of the disease.

References

Albrecht et al, 1998; Greenlee, 1998; Assensi et al, 1999; Clifford et al, 1999; Yiannoutsos et al, 1999; Tassie et al, 1999; Antinori et al, 2001; Cinque et al, 2001; Du Pasquier et al, 2001; Geschwind et al, 2001; Berenquer et al, 2003; Ferrante et al, 2003; Gray et al, 2003; Hoffmann et al, 2003; Dang et al, 2005.

Prognostic Factors

Severe immunosuppression creates conditions leading to the to PML development and is a predictive factor for a poor outcome of the disease. JCV-specific CD4+ T cells play a critical role in the control of JCV infection, prevention of PML, and also have a prognostic significance. The baseline CD4+ cell count greater than or equal to 100 cells/mm^3 is predictive for a prolonged survival; the CD4+ cell count of fewer than 100 cells/mm^3 is associated with an increased mortality.

It is well known that CD8+ T cells are important in the restraining of intracellular viral infection and eliminating of virus-infected cells. A number of studies confirmed that the specific cytotoxic CD8+ T cell response was pivotal in the suppressing of JCV infection and was associated with a favorable prognosis in HIV-positive PML patients. Early detection of JCV-specific cytotoxic T lymphocytes (CTLs) had an 87% predictive value for subsequent control of PML, while the absence of such CTLs had an 82% predictive value for subsequent active PML (p=0.0009). In reality, the patient evaluations, which took place less than four months after PML onset, showed that 78% of the patients with JCV-specific CTLs demonstrated subsequent disease control, whereas 100% of the patients without JCV-specific CTLs developed progressive PML (p = 0.007).

The concentration of JCV in the CSF is also a significant prognostic marker predictive for the course of PML. The JCV DNA concentration between 50 and 100 copies per microliter signifies whether the disease course will be moderate or severe. The patients with a JCV load below this level survive longer than those with a JCV load above this threshold.

References

von Einsiedel et al, 1993; Berger et al, 1998; Yiannoutsos et al, 1999; Antinori et al, 2001; Du Pasquier et al, 2001; Weber et al, 2001; Koralik et al, 2002; Berengeuer et al, 2003; Berger, 2003; Cinque et al, 2003; Fazakerley and Walker, 2003; Gasnault et al, 2003; Du Pasquier et al, 2004; Eash et al, 2004; Katz-Brull et al, 2004.

Key points:

1. The low incidence of PML in immunocompetent individuals and elevated rates of PML in immuno-compromised patients directly point to the crucial role of immunosuppression in the development of PML.

2. The introduction of HAART had little impact on the incidence of PML in HIV+/AIDS patients but prolonged survival of patients with HIV+/AIDS-associated PML.

3. The predictors of long-term survival in PML patients include the critical protective levels of CD4+ T cells and CD8+ T cells, a low JCV load, neurological recovery, and radiographic improvement.

The JC virus is the etiologic agent of PML. Following chapter presents the brief general information about viruses, modes of JC virus life cycle, virus-induced immunity, and causes of immunosuppresion that may lead to JC virus activation.

Chapter 5

Viruses

Virus life cycle

Viruses are obligate intracellular parasites that require a living host cell to perform virtually all virus biological functions. Viruses hijack the host's basic machinery utilizing it for the most aspects of the reproduction and multiplication of new viruses. One viral particle that infects a single host cell can produce thousands of progeny in the infected cell.

A virus has a small genome composed of a single nucleic acid—DNA or RNA. A third group of viruses uses both DNA and RNA as their genetic material but at different stages of their reproductive cycle. The virus genome is packaged in a protein coat (capsid), which in some viruses is further enclosed in a lipid bi-layer membrane envelope (**Figure** 7).

The major criteria used in the classification of viruses include the type of nucleic acid and the presence or absence of the lipid bi-layer membrane, or envelope. Additional distinctions of virus types include a single-stranded (ss) versus double-stranded (ds) DNA or RNA. Viruses can also be classified on the basis of the host they infect—animal viruses, plant viruses, and bacterial viruses.

Virus

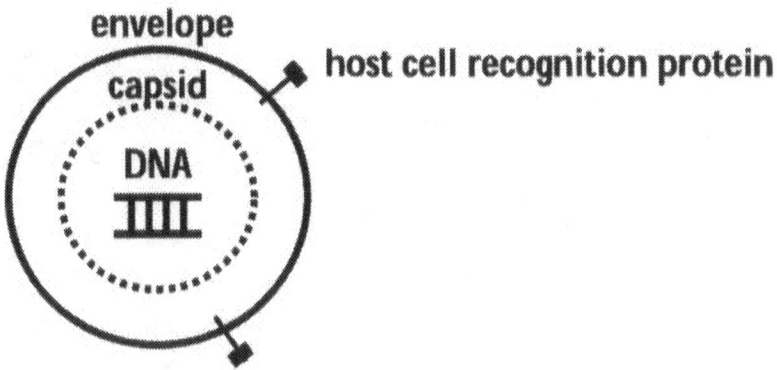

Figure 7. Schematic drawing of hypothetical double-stranded DNA, enveloped virus

Viruses have varied effects on the host cells. A lytic infection results in the destruction of the host cells, which allows the virus progeny to access the nearby cells. Many of the clinical manifestations of viral infection reflect this cytopathic effect of the viruses. Certain viruses may also cause latent infection of a host. In latent infection, there is a delay between the time of the cell infection by the virus and the appearance of symptoms as the virus emerges from latency under the influence of various stimuli. In persistent viral infection, the cells remain alive facilitating viral DNA replication over a long period of time in infected cells. A number of animal viruses have the potential for the transformation of a host cell from a normal cell to a cancer cell; in fact, the persistent viral infection is estimated to be the root cause of up to 20% of human malignancies (**Figure 8**).

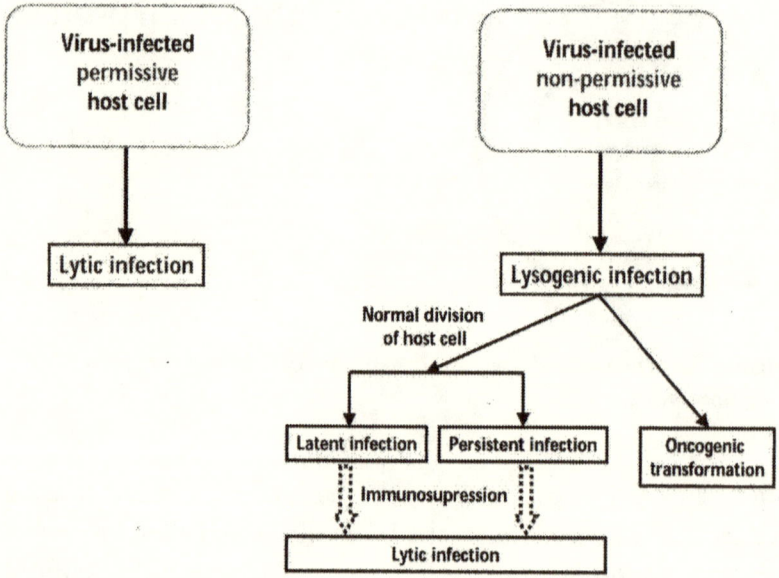

Figure 8. Possible consequences of virus infection
Virus infection may lead to the different consequences. Lytic infection in the productively infected permissive cells results in the virus replication, release of mature virions, and death of the host cells. Lysogenic infection in non-permissive cells may cause latent infection, persistent infection, or cell oncogenic transformation. Latent and persistent viral infections under some conditions, such as immunosuppression, can turn into a lytic infection.

The effect of a virus on the host cells primarily depends on the host cell species and phenotype, which fall into two classes: permissive and non-permissive cells. Permissive cells are productively infected: the viral cycle ends with the release of virus progeny and ultimately with the host cell death. Non-permissive cells cannot be infected productively, and viral replication is aborted. Some non-permissive virus-infected cells are transformed into malignant cells.

The lytic cycle of virus can be categorized into several steps (**Figure 9**). The first step is attachment. The virus binds to the surface of the host target cell. Binding is accomplished through the association of a viral surface protein with a specific receptor on the susceptible target cell surface. The second step is penetration. The virus enters the host cell and releases its genome from its protein coat or lipid envelope. Next intracellular stages of the virus life cycle include replication of the virus genome, synthesis of viral proteins necessary to create the capsid and envelope, and the assembly and packaging of the viral particles known as virions. The last step in the viral reproductive cycle is the release of mature virions from the host cell leading, in most cases, to the death of the host cell.

Lytic viral infection can lead to the host cell death via necrotic or apoptotic pathways. It is generally considered that upon the infection, most virus-infected cells initiate apoptosis as an altruistic response to curtail virus dissemination. Although it may seem detrimental to the host to destroy its own cells, such elimination of the infected cells actually spares the surrounding cells protecting them from the spread of virus and effectively restraining the viral infection. However, it has been suggested that viruses may be evolving to produce the specific agents or trigger signals that will enable them to evade this defense mechanism. Virus-infected mature cells of the central nervous system (CNS) are more refractory to apoptosis than are cells in other tissues. This relative resistance to apoptosis may be a major factor allowing many viruses to persist in the CNS cells.

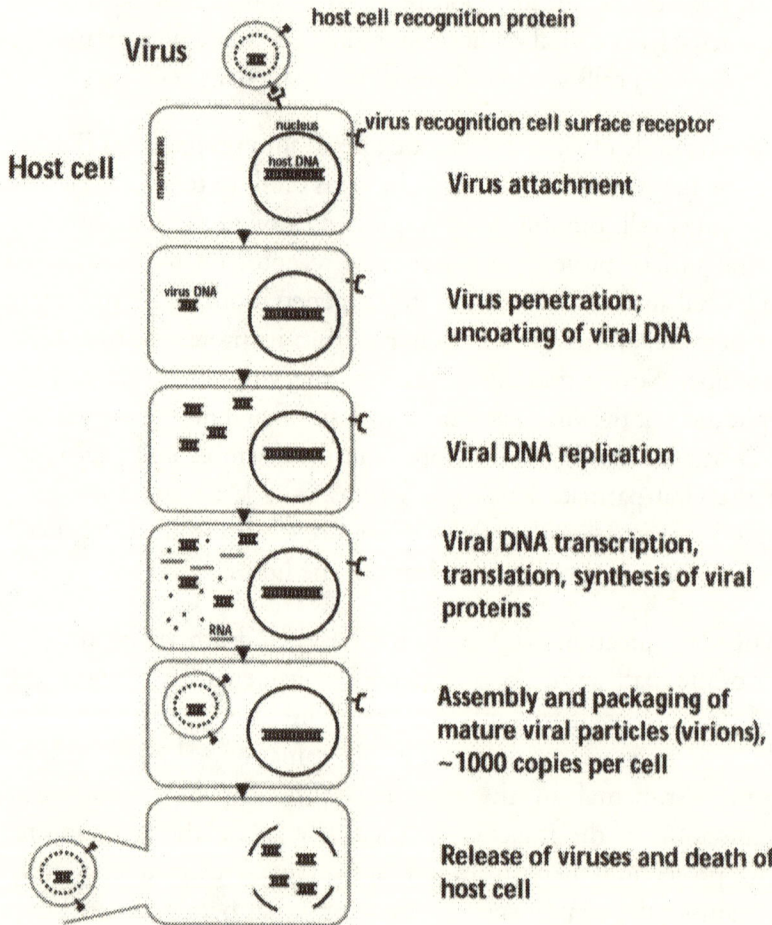

Figure 9. Cascade of virus lytic cycle

Refernces

Magigan et al, 1997; Alberts et al, 2002; Fazakerley and Walker R, 2003; Seth et al, 2004; Wang and Kieff, 2005.

Key Points:

1. Viruses require the living host cells in order to multiply and produce new viruses.

2. There are several types of viral infection: lytic, latent, persistent, and transformative.

3. The effect of a virus on the host cells primarily depends on the host cell species: permissive and non-permissive cells.

4. In most cases, death of virus-infected cells represents a host protective mechanism.

5. Critical steps in viral replication are performed by the host cell's own machinery. The development of effective antiviral drugs is particularly challenging because virus becomes a functional part of the host cells into which it is integrated, and it is difficult to identify the agents that can selectively attack the viruses without simultaneously destroying the host cells.

Virus-induced immunity

Infection with the most pathogens does not result in death of a host or even in the development of a disease. The body natural immunity eliminates the majority of viral infections. The immune system is the basis of the viral disease resistance. The immune mechanisms involve specific and nonspecific immune responses to virus invasion. Specific mechanisms include antibodies (Abs) and T cells, and non-specific mechanisms involve natural killer (NK) cells and interferons (IFNs). The immune system attacks viruses during both extra- and intracellular phases of their cycle.

Anti-virus antibodies bind directly to the intact viruses (**Figure 10**). This results in the neutralization of viruses. Antibody binding renders the viral pathogen incapable of infecting the host cells by preventing the attachment and penetration of virus. Antibodies inactivate or eliminate only extracellular pathogens, while intracellular pathogens are sheltered from the antibody attacks.

Whereas B cells recognize intact extracellular antigens and produce Abs, which inactivate only extracellular viruses, cytotoxic T lymphocytes (CTLs) recognize viruses inside of the host cells and directly kill the virus-infected cells (**Figure 11**). Actually, CTLs are the principal effector cells for clearing of established viral infections. Relative balance between virus spread and CTL response defines whether the virus will be eliminated or continue to persist. It was indicated that CTLs recognize degraded fragments of the internal viral proteins, which are bound to the class I MHC proteins on the surface of infected cell. The resulting complex—host I MHC proteins/viral proteins—is then attacked by the CTLs. Because class I MHC proteins are expressed on the surface of virtually all nucleated cells, effector CTLs are able to recognize any cells in the body that happen to become infected with the intracellular viruses. The presence of the above mentioned protein complexes in high concentrations activate the killing function of CTL cells. Many viruses have developed mechanisms to inhibit the high expression of class 1 MHC molecules on the surface of the cells they infect in order to avoid detection by the cytotoxic T lymphocytes.

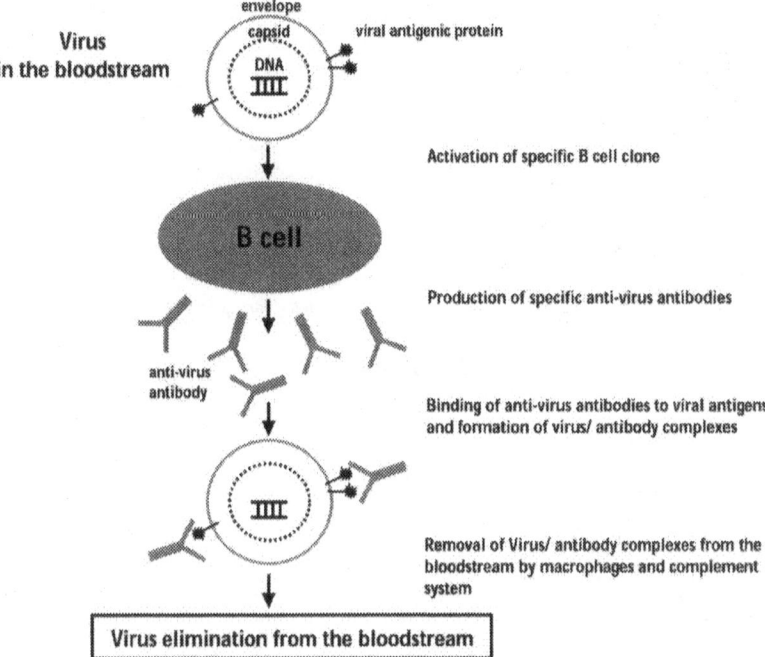

Figure 10. Humoral immune response to the viral invasion. Extracellular stage of virus infection

Virus-activated B cells produce antibodies against viral antigenic proteins, which bind the intact viral antigens. Viral antigen/antibody complex is a target for the phagocytosis by macrophages and neutralization by the complement system. These processes result in the virus elimination from the bloodstream and prevention of virus attachment and penetration into the host cells.

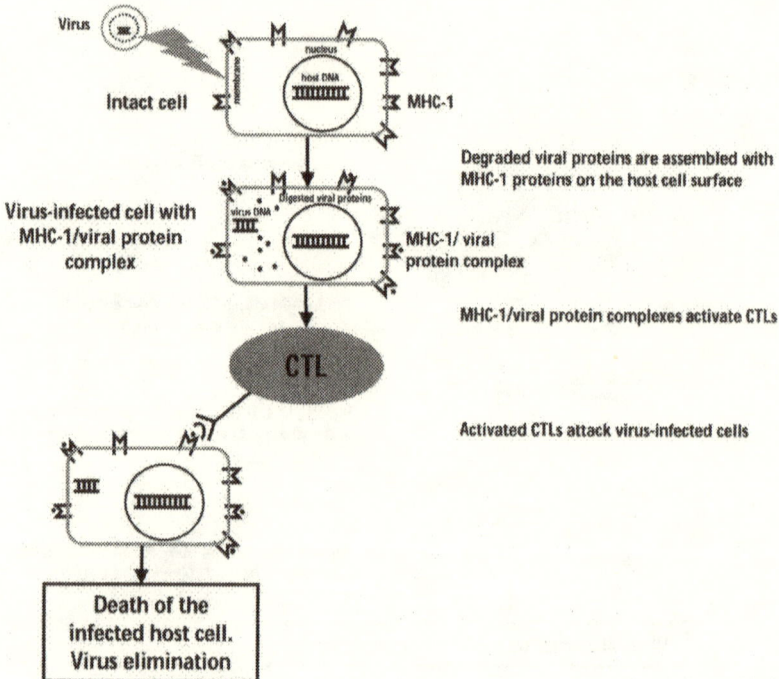

Figure 11. Cytotoxic T lymphocyte (CTL) response to the viral invasion. Intracellular stage of virus infection

CTLs are the key immune cells for the clearing of the intracellular virus infection. Activated CTLs bind to the MHC-1/viral peptide complexes assembled on the host cell surface, which lead to the death of virus-infected cells and virus elimination.

But viruses cannot successfully rescue themselves even by decreasing the expression of class 1 MHC molecules on the surface of infected cells because then they become victims of the activated NK cells (**Figure 12**). NK cells monitor the levels of class 1 MHC proteins. Low levels of class I MHC proteins activate killing function of NK cells. NK cells selectively destroy cells with low levels of class I MHC proteins including both virus-infected cells and some cancer cells. For this reason, it is difficult or even impossible for viruses to hide from both the adaptive CTL-mediated immunity and innate NK-mediated immunity simultaneously. Both NK cells and CTLs kill the virus-infected target cells by inducing them to undergo apoptosis before the virus has a chance to replicate. It is not surprising then that many viruses have acquired mechanisms to inhibit apoptosis, particularly, at early stages of the infection.

The antiviral function of IFNs is twofold. First, IFNs impede viral replication, a function that is critical for the host survival in response to viral infection. Second, IFNs stimulate both the adaptive and innate cellular immune responses. All of the machinery required for presenting viral antigens to CTLs is coordinated and called into action by IFNs via a positive feedback loop, which amplifies the immune response and culminates in the death of the virus-infected cells.

Failure of any of the immune system components severely jeopardizes the integrity of the immune system, leads to immunosuppression, and reduces an appropriate antiviral immune response.

References

Alberts et al, 2002; Alan and Gorska, 2003; Eash et al, 2004; Haynes and Fauci, 2005.

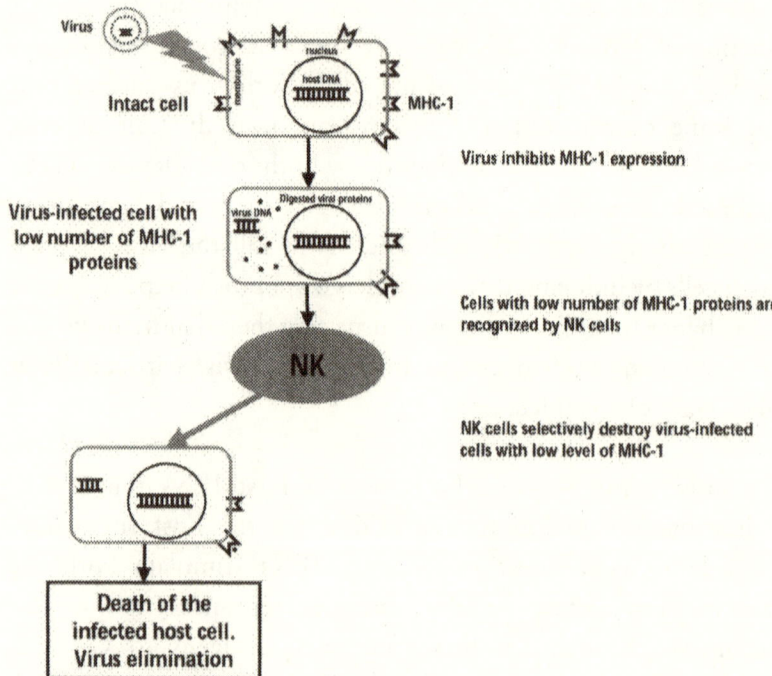

Figure 12. Natural killer (NK) cell response to the viral invasion. Intracellular stage of virus infection

Many viruses decrease expression of MHC-1 proteins on the surface of virus-infected cells. NK cells recognize and selectively destroy the cells with low concentration of the MHC-1, and thereby they eliminate viruses.

JC Virus

Structure

JCV is a small dsDNA, nonenveloped virus belonging to the family *Papoviridae* of the subfamily *Polyomavirinae*.

The viral genome consists of a closed circular single molecule of dsDNA, which is divided into three functional regions (**Figure 13**):

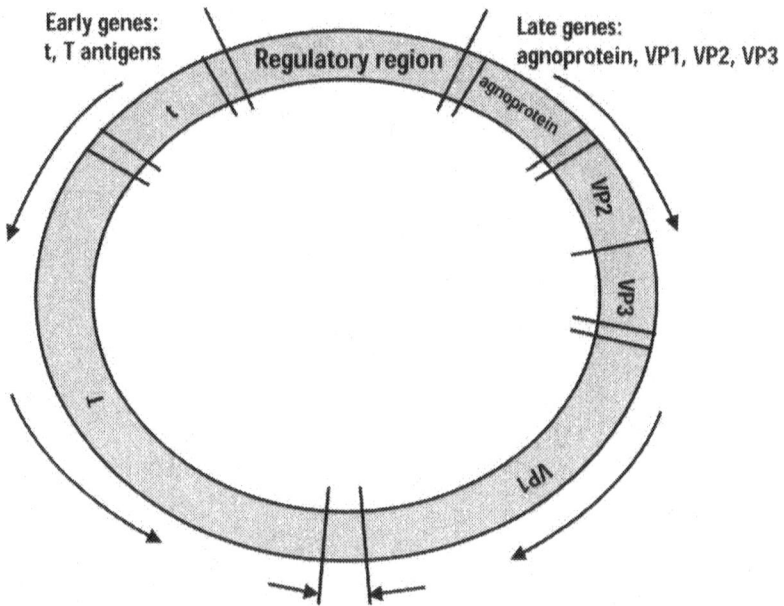

Figure 13. Schematic drawing of JCV genome structure
JCV genome is composed of the regulatory and coding regions. Coding region is divided into region coding the early genes (T and t antigens), and region coding the late genes (viral structural proteins—agnoprotein, VP1, VP2, and VP3).

1. The regulatory region, which contains the origin of DNA replication site and variable promoter and enhancer sequences. The regulatory region controls the viral DNA replication and transcription of early and late virus genes.

2. The early region is transcribed before DNA replication and encodes the large T antigen and the small t antigen proteins. The large T antigen and the small t antigen regulate transcription of the late region.

3. The late region is transcribed after DNA replication and encodes the structural capsid proteins VP1 (the major structural viral protein), VP2, VP3, and agnoprotein.

JCV regulatory region was discovered to exist in two major configurations—"archetype," which was commonly detected in the healthy immunocompetent people; and "PML-type," which was isolated from the brain tissue specimens of PML patients and characterized by a rearrangement of regulatory sequences.

Genotype Classification

Based on the sequence analysis of the complete genome and analysis of large T antigen and VP1 gene segments, the JC virus was classified into several major geographically specific genetic types: type 1, which is of European origin; type 2, which is Asian; and types 3 and 6, which are African in origin. Type 4 appears to be a recombinant type of the European type 1 and African type 3 versions, and this type is prevalent within the USA among African-Americans. It was suggested that JCV genotypes are determined by the geographical origin of the ethnic group rather than by the JCV genotypes that are dominant in the current location of the population. For example, although the slave trade in the USA ended in 1808, the ancient African JCV strain type 3, which existed in Africa 200 to 400 years ago, still persists in modern African-Americans. Studies of the worldwide distribution of the identified JCV genotypes are still underway.

The process of viral evolution is dynamic. Mixed human populations provide a locus for the emergence of more or less virulent strains. It was suggested that different spectra of the genotypes or genomic rearrangements of JCV may potentially vary in the virulence and in the clinical course of PML. To some extent, this idea is supported by the epidemiological data showing that PML occurs in 4–10% of AIDS patients in the USA and Europe; in 1.5% of AIDS patients in West Africa; and only in 0.8% of Brazilian AIDS patients. It is easy to speculate that higher incidence of AIDS-associated PML in the United States and Europe may be partly due to the virulent genotype of JC virus. Among AIDS/PML patients who received highly active antiviral treatment (HAART), 52% and 44% of the patients had JCV type 1 and type 2, respectively. Among HAART-treated AIDS/PML patients, 59% had JCV type 1, 23% type 2, and 18% type 4. Ferrante and co-authors suggested that the introduction of HAART may have contributed to changes in the distribution of the JCV genotypes. Italian study revealed a high frequency of dual infection with both JCV types 1 and 2 in PML patients. Dual infection with JVC 1 and 2 types might increase risk for the PML development.

References

Agostini et al, 1997; Agostini et al, 1998; Chima et al, 1998; Caldarelli-Stefano et al, 1999; Chima et al, 1999; Bofill-Mas et al, 2000; Ferrante et al, 2001; Del Valle et al, 2002; Sariyer et al, 2004.

JCV Receptors

As for all viruses, the life cycle of human JCV begins with the virus attachment to the surface of target cells. The primary interaction between the virus and the target cells is achieved by the attachment of the viral particles to the host's cell surface receptors. Human JCV specifically binds to the cell surface glycoprotein receptors containing α-(2-6)-linked sialic acids. The initial virus-host cell interaction is considered to be a major determinant of the viruses' host range and tissue tropism. In the brain, JCV sialic acid receptors are expressed on oligodendrocytes and astrocytes, but not on cortical neurons, which underlies the selective JCV tropism to CNS glial cells. Also high levels of glycoprotein receptors containing sialic acids are present on the surface of extra-neural cells such as B lymphocytes, tonsil, spleen, kidney, and lung cells. After viral internalization, the serotonergic receptor 5HT2AR acts as a cellular receptor for JCV on the human glial cells.

Experiments with ds-DNA murine *Pyomavirus* showed that the initial interaction of the murine *Pyomavirus* with the host cells occurs through direct binding of the major capsid protein VP1 with the sialic acid cell receptor. Data on virus infectivity indicated that $\alpha_4\beta_1$ integrin turned out to be a cell receptor for the murine *Pyomavirus* A2 strain. It is likely that $\alpha_4\beta_1$ integrin is involved in virus-cell interactions in a post-attachment stage.

References

Pho et al, 2000; Caruso et al, 2003; Eash et al, 2004; Elphick et al, 2004.

JCV Life Cycle

Following the attachment, the JCV penetrates the plasma membrane and transports its genome to the nucleus of the host's cells. After this stage, JCV has three extremely different life modes: the persistent nonpathogenic asymptomatic mode; the severely pathogenic mode that induces PML; and, recently discovered, third mode that is associated with cancer transformation.

Persistent Nonpathogenic Infection

JCV is a common ubiquitous human virus, nonpathogenic in the majority of cases. The immuno-epidemiological studies revealed that most of the world's population has been exposed to JCV. By the age of ten, 40–60% of the total population is JCV seropositive. By adulthood, this percentage is even higher with seroconversion rates exceeding 80% of the population worldwide and 90% in some urban areas.

After initial infection of the host cells, the viruses establish a life-long persistent asymptomatic infection in healthy individuals. The virus dwells in a wide range of human cells such as B-lymphocytes, tonsil, kidney, lung, and gastrointestinal tract cells. More than 50% of healthy adults harbor JCV DNA in their kidneys. JCV is frequently excreted in urine without presenting any apparent clinical symptoms. Most likely that intermittent viruria results from the activation of a previously established JCV infection in the urinary tract. This signifies that renal JCV is not completely latent but can replicate and generate progeny. In healthy immuno-competent individuals the rate of renal JCV excretion increases with age.

Reference

Bofill-Mas et al, 2000; Ricciardiello et al, 2000; Ricciardiello et al, 2001; Berger, 2003; Fazakerley and Walker, 2003; Eash et al, 2004; Seth et al, 2004.

Lytic Pathogenic Infection—PML

In permissive microglial cells, JCV infection results in the formation of new virions and lytic destruction of the host's oligodendrocytes and astrocytes. The lytic pathogenic form of JCV has restricted neurotropism. The glial brain cells are the only known principal targets of JCV active virulent infection with subsequent PML manifestations. Currently, it is unclear whether the provocative condition of the immunosuppression allows JCV replication outside of the brain followed by the penetration of active viruses into the brain, or JC viruses presents in the brain where actively multiply upon the immunosuppression and then spread to the various regions within the brain. Experiments with human astrocytes demonstrated that JCV-induced astrocyte death results from necrosis, not apoptosis. However, apoptosis cannot be completely eliminated as a mechanism of oligodendrocyte death.

Virulent JCV genotype (PML type) in the brain is different from the nonpathogenic JCV genotype (archetype) in the kidney, lung, and B cells. In PML type of JCV altered viral DNA regulatory region is formed from the archetypal JCV regulatory region via unique rearrangements by the deletions and duplications within JCV promoter and enhancer. Consequently, the formerly nonpathogenic virus is transformed into the pathogenic PML causative type capable to replicate in the glial cells and to lyse them. Despite the suggestion that the determinants of JCV neurotropism and neurovirulence are located in the viral DNA regulatory region, experiments failed to reveal any PML specific genetic markers.

A population based analysis showed very high incidence of PML among the patients with AIDS compared to the patients with other condition with immunosuppression, for example, 5.1% of AIDS patients versus 0.07% of the patients with hematologic malignancies. This fact implies that in addition to the profound effect of HIV on T cell-mediated immunity, it is highly likely that other factors specific to HIV contribute to the extraordinarily high frequency of PML occurring in the setting of HIV infection. This suggestion was confirmed by the experiments which showed that HIV-1 secreted proteins stimulated exponential replication of JCV by transactivation of the JCV late promoter. HIV gene products seemed to be able to translocate the JCV viral promoter directly. Hence, the high incidence of PML among AIDS patients could be explained by HIV-mediated immunosuppression as well as by the virus interference of HIV-1 and JCV.

Transformations of the archetypal JCV strain into the virulent form are apparently associated with the host genome regulatory mechanisms. Experiments with eukaryotic cell chromatin performed by Ermekova and co-workers suggested that genomic reprogramming (i.e., activation of inactive genes or repression of active genes) is under the control of cell chromatin tissue-specific proteins. These proteins localize on the eukaryotic cell DNA regulatory regions. Because all virus functions are maintained by the host's cell machinery, changes in viral genome activity may be regulated by the host's tissue-specific proteins, which determine the conversion of the nonpathogenic form into the pathogenic virus in the permissive host cells.

Transcriptional control region without rearrangements was detected in two PML long-term survivors after they underwent HAART therapy. This finding led to speculation that HAART may have reduced the rate of the virus transcriptional control region rearrangements in these two AIDS/PML patients.

References

Ermekova et al, 1984a; Ermekova et al, 1984b; Major et al, 1992; Agostini et al, 1998; Greenlee, 1998; Saito et al, 1998; Caldarelli-Stefano et al, 1999; Pho et al, 2000; Power et al, 2000; Dorries, 2001; Berger, 2003; Fazakerley and Walker R, 2003; Ferrante et al, 2003; Daniel et al, 2004; Eash et al, 2004; Enam et al, 2004; Seth et al, 2004; Zheng et al, 2004; Dang et al, 2005.

Oncogenic Transformation

Association of JCV with Experimental and Human Tumors

Initiation of the JCV lytic cycle in the glial cells is a pivotal event in the pathogenesis of PML. At the same time, it was demonstrated that JCV induced neoplastic transformation in cell cultures. It was also shown that inoculation of JCV into several experimental animal models, including nonhuman primates, resulted in the formation of a wide variety of brain tumors such as medulloblastomas, astrocytomas, and glioblastomas. Evidently, JCV lytic infection results in destruction of cells, whereas JCV abortive replication may cause cell transformation.

A group of researchers from the Temple University found a strong statistical correlation between JCV and the development of a broad range of human brain tumors. Intensive analysis performed by Dr. Del Valle with co-workers showed that 83.3% of ependimomas, 80% of pilocytic astrocytomas, 76.9% of astrocytomas, 66% of anaplastic oligodendrogliomas, 62.5% of oligoastrocytomas, and 57.1% of oligodendrogliomas contained JCV early gene sequences. Recently the JCV genome was found in colorectal cancers. The presence of the JCV genome was determined in precancerous villous adenomas. The fact that the JCV genome is identified in such a wide variety of human tumors raises the suggestion that JCV plays an important role in human tumorogenesis.

Expression of large T antigen was found in a variety of human tumors. At the same time, expression of the JCV late gene product, VP1 capsid protein, was not detected in any tumor types. These observations provide a suggestive evidence for a possible association of early and late JCV proteins with the neoplastic cell transformation and uncontrolled cell proliferation.

PML and Tumors

Reiss and Khalili and later Sariyer with co-authors reviewed several published cases of the coincidence of PML occurrences along with different tumors. In 1961, Richardson described the case of the patient with chronic lymphocytic leukemia and PML in whom postmortem examination revealed the presence of an oligodendroglioma. In 1974, Castenge and co-workers decribed the case of patient who had a long history of immunodeficiency syndrome, died from PML and in whom postmortem examination detected numerous of anaplastic astrocytes. In 1983, Sima and others observed the patient with PML and multiple astrocytomas. In the last two cases, microscopic analysis of the demyelinating lesions demonstrated the presence of viral particles in both oligodendrocytes and astrocytes within PML foci, but not in the neoplastic astrocytes. In 2000, Shintaki and colleagues reported the case of PML in association with dysplastic ganglion-like cells. Neurons that were infected with JCV expressed large T antigen in the absence of capsid protein, VP1.

Astrocytoma or PML?

Van Assche and co-workers described the case of Chron's disease patient who participated in ENACT-1/ENACT-2 trials, received eight doses of natalizumab, and died during ENACT-2 trial with diagnosis astrocytoma. Two years later, the patient's case was repeatedly reviewed, and the initial diagnosis of astrocytoma was reclassified as JCV-related PML. But in the light of mentioned above observations about concomitant occurrences of PML and brain tumors, the possibility of double diagnosis for this patient has to be discussed. For your judgment, the detailed patient's catamnesis presented below.

In July of 2003, patient, who received eight monthly natalizumab doses for treatment of Crohn's disease, was admitted to the emergency unit with the severe neurological and mental symptoms. Trephination was performed with partial resection of the right frontal lesion. Based on the clinical picture, the large size of the frontal lesion, a high number of astrocytes with very large and atypical nuclei, and high Ki-67-MiB1 proliferation index (± 15%) the diagnosis of grade III astrocytoma was made (according to the WHO criteria). The patient's condition deteriorated rapidly. Patient died in December 2003, five months after admission. Autopsy was not performed.

After the Elan Pharmaceuticals and Biogen Idec announcement in February, 2005, regarding the occurrence of PML in two natalizumab-treated MS patients, this Crohn's disease patient's case was repeatedly reviewed, and additional tests with frozen blood serum samples and brain tissue specimens were performed. The retrospective reappraisal of the patient's clinical course indicated that the initial symptoms were suggestive for PML. PCR detected JCV DNA in the frozen blood serum samples, which were obtained two months before the admission; JCV DNA concentration increased by a factor of 12 by the time the patient was admitted. Histological reexamination of the brain specimens demonstrated changes in oligodendrocytes suggestive for PML; brain lesion tissues contained a high viral load; immunohistochemical assay with Mab directed against the SV40 large T antigen revealed the staining of atypical astrocyte's and oligodendrocyte's nuclei. Based on these new data, investigators reclassified the previous diagnosis of astrocytoma to the diagnosis of PML. Authors concluded that the temporal correlation between monthly natalizumab treatments and the increase in levels of JCV replication clearly illustrated the existence of the direct relationship between natalizumab therapy and PML development.

At the same time, it seems that in this particular case a correct diagnosis of PML does not exclude the possibility of a correct diagnosis of astrocytoma. Actually, the investigator's first pathology report has provided an extremely strong evidence for the existence of astrocytoma. The most convincing indications for the diagnosis of tumor include presence of astrocytes with atypical morphology and a high Ki67-MiB1 proliferation index. Ki67 is the prototypic cell cycle-related nuclear protein expressed only by proliferating cells at all phases of the active cell cycle (G-1, S, G-2, and M phase), but not in the resting cells (G0 phase). Immunohistochemistry with Mab Ki67 evaluates the rate of cell proliferation. Ki67-MiB1 proliferation index detects the cell growth fraction by calculating the percentage of Ki67 positive cells among the total number of cells. The positive correlation between the proliferation marker and neoplasms is very strong. Moreover, because normal brain cells have very low levels of proliferation, Ki67-MiB1 proliferation index in the brain equal to or greater than 15% is highly suggestive for the brain tumors.

The positive immunostaining for the large T antigen during reexamination of the brain specimens is indicative for both diagnoses—PML and cancer—moreover, with a suggestive prevalence for the cancer diagnosis. Although, it was suggested that T antigen may directly alter expression of the myelin proteins or may inhibit the maturation of oligodendrocytes, which indirectly alter the myelin layer around the axons and lead to the secondary demyelination, it is well proven that JCV large T antigen is the main viral regulatory protein that is involved in the tumor induction (**Figure 14**). Thus, in this particular case it is possible to believe that both diagnoses were correct and the patient had concurrent astrocytoma and PML.

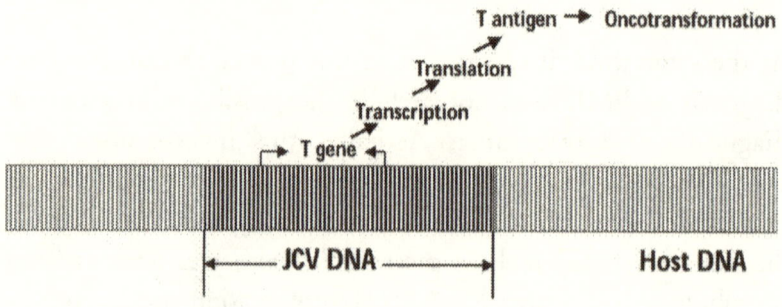

Figure 14. Schematic drawing of hypothetical scheme of cell oncotransformation by JCV
JCV DNA is randomly incorporated into the host cell DNA. In the presence of specific host gene regulatory proteins, the JCV early genes are transcribed (T antigen) that can transform host cells into the cancer cells.

Possible Mechanisms of JCV Involvement in Cancer Development

- It is well established that the inactivation of tumor suppressor proteins such as p53 and pRb is a crucial step in the cascade of events leading to tumorogenesis. The oncogenic capacity of T antigen of polyomaviruses including JCV is also proven. Upon binding to p53 or pRb, T antigen inactivates tumor suppressor proteins p53 and pRb, dysregulates several proliferation pathways, and interferes with the cell cycle progression.

- It is demonstrated that agnoprotein plays a critical role in the replication of JCV. In the absence of a viral lytic infection, stimulation of cellular events by agnoprotein may lead to the rapid uncontrolled cell growth.

- It is known that the absence of the viral capsid proteins excludes the productive replication and leads to the viral abortive cycle. The absence of VP1 capsid protein in neurons containing JCV and expressing JCV T antigen suggested that these cells are not productively infected with JCV, but could have been transformed by the virus.

References

Del Valle et al, 2002; Enam et al, 2002; Reiss and Khalili, 2003, Sariyer et al, 2004; Van Assche et al, 2005.

Key points:

1. Data from the cell cultures and animal experimental models provided an evidence of the involvement of JCV in the process of cellular transformation and tumorogenesis.

2. Detection of the surprisingly high amount of JCV DNA sequences and viral proteins—large T antigen and agnoprotein—in a variety of human tumor cells strongly suggests a possibility that this virus may play a role in the development of human cancers. However, these data do not clarify the mechanisms by which JCV induces tumors in humans.

3. It is hypothesized that JCV exerts its oncogenic activity, at least in part, through the early proteins of the virus, particularly T antigen, by the association with several critical tumor suppressor cellular proteins—p53 and members of the pRB family. The inactivation of tumor suppressor proteins of the host by viral proteins leads to a deregulation of cellular growth and uncontrolled proliferation.

4. Although the precise mechanisms of JCV activity in human tumorogenesis has yet to be clarified, the presence of JCV genome and, more importantly, the expression of its proteins in the absence of productive replication suggest the potential involvement of JCV, possibly as a co-factor, in the development and/or the progression of cancer.

Chapter 6

Summary and Reflections

Summary

Topographic and functional diversity of α_4 integrins

Direct contact of cells to other cells and to the extracellular matrix (ECM) exerts a profound control over cell functions and is required for practically all cell operations. Cells adhere to adjacent cells and to the ECM via adhesion molecules. Integrins—transmembrane heterodimeric receptors—belong to the group of adhesion proteins and by participating in cell anchorage play a critical role in a number of processes on the systemic, tissue, cell, and molecular levels. These include immunity; homeostasis and hematopoiesis; embryogenesis and fertility; establishment and maintenance of tissue architecture; cell proliferation and differentiation, cell cycle division, and gene expression. The integrin's involvement in so broad range of biological events can be explained by two major reasons—the ability of integrins to serve as cell bidirectional signaling molecules and by their widespread locations.

Serving as bidirectional signaling molecules, integrins provide inside-out and outside-in signaling mechanisms, which could be considered as two-step regulation of multiple cell processes. On the first step (inside-out), intracellular signals, from integrin-bearing cells primary activated by external or internal stimuli, elicit conformational changes in the integrins that enhance the integrin's binding properties. Inside-out signaling is mostly responsible for the adhesion and migration of cells to the sites of demand. One of the clearest examples of the cascade of inside-out signaling is immune cell trafficking to the inflammatory site, which includes the following phases: the initial activation of leukocytes by specific inflammatory soluble factors, the transmission of intracellular signals to the integrins on the leukocyte's surface, the switching of integrins to an adhesion-competent state, the integrin-ligand binding with resultant firm adhesion of the rolling leukocytes to the endothelium, and finally the transendothelial migration of leukocytes into the extravascular area. The second step of the integrin's bidirectional signaling (outside-in) is initiated by integrin binding to the ligands. The ligation of integrins induces signals that are transmitted back to the cells where they modify a large variety of intracellular reactions. Outside-in signaling coordinates survival/apoptosis, cell cycle division, cell proliferation and differentiation, and gene expression. Inside-out and outside-in mechanisms are closely linked through the integrin-ligand binding stage. Integrin-ligand binding provides signals to direct the cellular traffic to the sites of demand, on the one hand, and to control cell molecular transduction pathways, on the other.

Important role of integrins in the body is accentuated by the fact that almost all cells of the body (at various stages of differentiation—from stem and progenitor cells to the mature function-specific competent cells) express integrins. The widespread locations of integrins supports the suggestion that integrins are one of those universal elements that are critical for maintaning the basic processes in various tissues of the body. At least three of the body's systems—immune, blood, and reproductive—have now been shown to be essentially dependent in their functioning on the integrin activity.

Immune system

Integrins of α_4 subfamily are present on the surface of all mature cells of the immune system. Providing cell-to-cell and cell-to-EMC adhesion, α_4 integrins participate in a broad spectrum of immune responses to pathogens.

Despite the existence of numerous safe pharmacological agents that are effective against virulent organisms, infectious diseases remain a major cause of death and debility for people around the world. Moreover, in the last decade, mortality from infectious diseases has increased. The current danger of infections—the renewed intensity of old diseases and the appearance of new clinical entities—is due to several reasons, which include, but are not limited to: drug-resistant strains among all classes of mammalian pathogens, mutations in human viruses; the movement of animal viruses from their niche to become hazardous human pathogens; and immune status changes in certain groups of patients, including patients with secondary HIV-immunodeficiency and patients who are treated with pharmacological agents leading to immunosuppression—organ transplant recipients, oncology patients, and patients with inflammatory and autoimmune diseases. Individuals with decreased immune activity form a major proportion of the patients in whom not only primary pathogens but even opportunistic microorganisms often induce the development of severe infectious diseases.

Modern medicine has at its disposal very potential etiological drugs for the treatment of diseases and the complications induced by bacteria, fungi, parasites and to a much lesser extend viral diseases. At the same time, prevention/prophylaxis of these disorders is completely conditioned by immune defenses of the body—natural immunity or artificially-induced immunity (active/passive immunization). In the regard to virus-induced diseases, which are one of the greatest persistent unsurmountable perils in the modern world, the strengthening of body immunity is not only the best but actually the only strategy for disease prevention, as well as for a better prognosis for a disease that is already established.

In immunocompetent individuals invasion by opportunistic pathogens, usually, does not result in the development of the disease because the body immune system neutralizes or abolishes the foreign agents. The basis of resistance to viruses or the success of a natural (self-healing) curative process of viral diseases are depend on the person's immune attack on the viruses during extra- and intracellular phases of infection. B cell-produced specific antibodies, the complement system, and phagocitic cells inactivate/eliminate extracellular intact viruses from the peripheral blood and restrain the intracellular infection of the host cells. In most cases, high levels of humoral immunity completely prevent the development of disease. If antibodies fail to preclude penetration of viruses into the host cells, the cytotoxic T lymphocytes (CTLs) disrupt the virus-infected cells on whose surface degraded viral proteins are presented by the host major histocompatibility molecules (MHC 1); the specificity of virus-infected cell recognition by CTLs is determined by the viral component of the complex, and the degree of the CTL's killing activity is defined by the concentration of MHC 1 proteins. Many viruses inhibit the expression of MHC 1 molecules on the surface of the cells they infect, which decreases the activity of CTLs but at the same time increases activity of natural killer (NK) cells. NK cells monitor the concentration of surface MHC 1 proteins and selectively destroy virus-infected cells bearing low levels of I MHC proteins, which have previously avoided destruction by CTLs. Thus, immunocompetent hosts have a series of humoral and cellular immune barricades to prevent the development of infectious disease: B cell-produced antibodies render viral pathogens incapable of infecting host cells; and the coordinated actions of CTL and NK cells clear the established intracellular viral infections by targeting and killing, by apoptosis, different assortments of virus-infected cells before the virus' massive intracellular replication.

[Note: Many more effectors—different cell types and soluble factors—of innate and adaptive immunity participate in immune protection against pathogens; for the sake of brevity and simplicity, they are not discussed in this brief overview of antiviral immune responses.]

One of the typical examples of immune status influence on the virus resistance and virus vulnerability is unequivocally exhibited by the epidemiology of JC virus-caused life-threatening disease—progressive multifocal leukoencephalopathy (PML). By adulthood, 80-90% of the total human population is JC virus seropositive, which definitely evidences of high incidence of virus invasion. At the same time, in immunocompetent individuals, the development of the disease is actually a virtual medical curiosity; but among immunocompromised people the chance of PML development is very high (0.3%–10%) and may correlate with degree of immunosuppression. Currently, in the USA and in other developed countries, HIV is the underlying condition in 85%-90% of PML cases; and 10%-15% of all PML is attributable to other immunosuppressive factors—such as therapy for cancer and autoimmune diseases.

Integrin-mediated cell adhesion is mandatory for immune cells to act to the fullest extent. Lymphocytes bearing integrins with diminished adhesive properties cannot effectively exterminate pathological agents. Changes in integrin-mediated cell adhesion severely jeopardize the natural immune system as well as the responsiveness to the active immunization. Therefore, the integrin's inadequacy, through the direct reduction of immune cell activity, impacts on the risk of infectious disease development after the host's exposure to potential pathogens or in the course of an established disease. Profound changes in integrin functions— inhibition of α_4 integrin's expression or blockage of expressed integrins—cause generalized immunosuppression, creating an extreme risk of very serious infectious consequences. The direct dependence of the body immune status and any consequent susceptibility to infections from the integrin's functionality require a great deal of caution before one uses integrin-antagonist drugs. Pharmacological immunosuppression, particularly prolonged therapy leading to profound immunosuppression, may promote severe infections and must be put into medical practice judiciously only as a rescue therapy.

Blood system

The majority of hematopoietic stem and progenitors cells (HSPCs) express integrins of α_4 subfamily. The abundance of constitutively active α_4 integrins on the surface of erythroid, myeloid, and lymphoid hematopoietic progenitors suggests that integrins are involved in the functioning of all classes of hematopoietic cells. The integration of integrin's bidirectional signaling networks is necessary for the regulation of both major processes in the blood system—homeostasis and hematopoiesis.

Inside-out signaling is mostly responsible for homeostasis. In a steady state, the proliferating hematopoietic cells in various stages of development are confined within specialized bone marrow (BM) "niches," whereas terminally differentiated mature cells leave the BM and migrate into the blood to meet physiological or urgent demands. Providing stable adhesion of HSPCs to adjacent stromal cells and extracellular matrix components, integrins regulate the cell's trafficking—migration and anchoring within the BM, retention of immature blood cells inside of the BM, and transmigration of mature cells into the peripheral circulation (mobilization). Outside-in integrin's signaling in HSPCs is of particular importance in hematopoiesis: integrin-ligand binding is critical to the regulation of promoting differentiation and inhibiting proliferation of HSPCs in the BM.

Long-term reduction of α_4 integrins$_1$ action could lead to abnormalities in the homeostasis. In physiological condition, hematopoietic cells express a diminished amount of α_4 integrins at the final stages of cell transmigration through the endothelial sinuses. Levels of adhesion-competent integrins on the hematopoietic cells' surface presumably selectively control the retention of immature cells with high adhesive ability within the BM and the mobilization of mature blood cells with decreased adhesive ability. This assumption was confirmed by the integrin functional down-regulation with the integrin antagonist: experiments with baboons and macaques showed that injection of saturating amounts of anti-$\alpha_4\beta_1$ intergin antibodies dramatically increased (up to 200-fold) recruitment into the bloodstream of all classes of progenitor cells, including progenitors with high self-renewal potential. It is crucial to note that from the clinical standpoint the increased mobilization of immature progenitors from the BM to the peripheral blood may have worrisome consequences.

Besides mobilization, α_4 integrin deficiency may affect hematopoiesis. Given that $\alpha_4\beta_1$ integrin-mediated adhesion of progenitor cells to bone marrow fibronectin results in proliferation-inhibitory signaling, it is possible that defective integrin-dependent adhesive interactions leading to continuous proliferation of hematopoietic cells may be implicated in the pathogenesis of hematopoietic abnormalities. Long-term blockage of α_4 integrin function by anti-$\alpha_4\beta_1$ antibodies, causing uncontrolled proliferation of hematopoietic cells without balanced development into terminally differentiated erythroid, lymphoid, and myeloid cells cell, raises theoretical concerns of an increased risk of malignancies.

Reproductive system

Integrins of α_4 subfamily are present on the surface of germ cells, mature mammalian gametes, and embryonic cells at all stages of embryonic development except during blastocyst outgrowth. Integrin-mediated cell adhesion plays an important role in the major reproductive processes—spermatogenesis/oogenesis, sperm-egg interaction, and embryogenesis.

Sperm-egg fusion is initiated primarily by integrin-ligand binding. A positive correlation has been identified between integrin presence on the surface of spermatozoa and successful fertilization. Compared to normal sperm, the sperm of infertile patients exhibits significantly reduced amount of integrins, which account for decreased sperm-to-egg adhesion and some forms of infertility. Embryonic cells express developmentally regulated complex repertoire of adhesion molecules. There is speculation that differential expression of various integrins at early stages of embryogenesis may contribute to the embryonic cell attachment and may be important for the regulation of embryo development, survival, and interrelations with the maternal environment, which is essential for establishing maternal tolerance of the fetal allograft. Apart from the adhesion function, integrin-ligand interactions mediate outside-in intracellular signaling, which induce substantial cytoskeletal reorganization, changes in the tyrosine phosphorylation of proteins, and the modulation of gene expression in germ and embryonic cells. Integrin impairment due to any cause, including inhibition of its functions by anti-α_4 integrin antibodies, may lead to alterations in spermatogenesis/oogenesis, reduced fertility, and defects in embryonic development.

In addition to the impact of the α_4 integrins on immunity, hematopoiesis, and reproduction, it was recently demonstrated that integrins are engaged in other physiological processes: adhesion-mediated microglial expression of inflammatory cytokines and modulation of vascular myogenic tone.

Thus, integrins participate in the regulation of a numerous of vital body phenomena. The molecular interactions responsible for cellular adhesion are complex, well orchestrated and under sophisticated control. Alteration in the integrin's signal transmission can lead to dangerous consequences. It is essential to remember that innovative integrin-antagonist therapy, which transforms the integral functioning of integrins, has to be utilized with extreme caution.

Specificity of monoclonal antibody therapy

Monoclonal antibodies (Mabs) are being increasingly used for therapeutic and diagnostic purposes and have a significant impact on many fields of medicine, especially oncology, transplantation, and inflammatory and autoimmune pathologies. The emergences of new recombinant technologies and molecular strategies has led to the manufacturing of expanded repertories of Mabs—chimeric, humanized, fully human, small fragment, and bispecific. Modern methods of monoclonal antibody production are so sophisticated that they appear to be more an art form than a mere technique. Monoclonal antibodies of almost any required design can be constructed to maintain the most crucial feature of antibodies— its specificity to target antigens, and this makes antibodies a highly potent antigen-specific tool.

However, antibodies represent only one partner of the "couple" in specific immunotherapy. The second partner is the antigen itself. The crux is that since the antibodies bound antigens, they neutralize/eliminate soluble antigens from the blood or interfere with the functioning of all antigen-bearing cells, whether indicated for therapeutic targeting or not. In order to realize the precise goal of an immunotherapy, the selected antigen must be strictly specific. Ideally, the cell's antigen has to be expressed only by the cells marked for functional inhibition or elinimation. It is not ideal but quite permissible, if the population of non-indicated cells expressing the antigen is limited to non-imperative cells. It is less desirable but in some situations still acceptable, if a restricted population of vitally important cells contain the selected antigen or express the antigen with much less intensity than therapeutically targeted cells. Inevitable destruction of all cells expressing antigens, including those that are not to be destroyed for therapeutic goal, has to be taken into account during the development of new therapeutic monoclonal antibodies and in any assessment of the benefit/risk ratio. Otherwise, power of the antibody may become a dangerous weapon.

Monoclonal antibody natalizimab (Tysabri), which is used against $\alpha_4\beta_1$ integrins, was developed and reported to be very effective for treatment of multiple sclerosis and Crohn's disease. It works by binding to $\alpha_4\beta_1$ integrins on the surface of immune T cells. Obviously, natalizimab (Tysabri) as an antibody is highly specific to α_4 integrins, but α_4 integrins, are not distinctive markers for immune T cells only. It is expressed by a broad range of immune, hematopoietic, germ, and embryonic cells; moreover, targeted mature immune T cells exhibit diminished levels of $\alpha_4\beta_1$ integrins compared to the hematopoietic stem and progenitor cells. Thus, treatment with natalizimab (Tysabri), as an immunotherapy, is not strictly specific to the targeted T cells. Theoretically, prolonged blockage of α_4 integrins by natalizimab (Tysabri) may impair integral functioning of the immune, hematopoietic, and reproductive systems.

Synopsis of trial data analysis with natalizumab (Tysabri)

Pioneer in the field of new class of drugs, integrin antagonists, is anti-α_4 integrin monoclonal antibody, natalizimab (Tysabri). Analysis of trials is based on all published data and results reported in the Elan Pharmaceutical and Biogen Idec press releases.

(Note: Detailed trial data analysis is presented in chapter III, "Anti-integrin monoclonal antibody treatment for multiple sclerosis and Crohn's disease")

Anti-α$_4$ integrin antibody effects on animal models of multiple sclerosis, EAE

1. Antibodies against α$_4$ integrin prevented the development of EAE.

2. Antibodies against α$_4$ integrin inhibited short-term progression of an established EAE but were much more effective in preventing EAE than in restraining an established disease.

3. For some models, antibodies against α$_4$ integrin increased the activity of ongoing EAE.

4. Antibodies against α$_4$ integrin were effective only during the treatment period.

5. After the treatment ended there was a rapid return to symptomatic EAE.

Table 1 lists all placebo-controlled trials that were performed to assess the efficacy and safety of natalizumab treatment for acute Crohn's disease and multiple sclerosis.

Crohn's disease				
Trial	Duration Patients (n) Infusions [n]	Efficacy (+ or -)	Side effects during/after treatment (no/yes/list)	Comments
UK	4 wks (30) [1]	+	no	The beneficial effect was short-lived, and rescue therapy was required for most patients at week 4
Europe Israel	6 wks (248) [2]	+	no	The beneficial effect was short-lived, and persisted only through week 12 of follow-up
ENACT 1	?	?	?	Trial data are not published
ENACT 2	?	?	One case of lethal PML associated with treatment	Trial data are not published. PML case is published

Multiple sclerosis				
Phase I	8wks (28) [1]	+	no	The authors concluded that a single infusion of natalizumab (0.01 to 3 mg/kg of body weight) is associated with extremely strong, long-lasting (one year) beneficial effect. These data differ from the results of other studies
UK Phase II	24 wks (72) [2]	+	Leukocytosis; antibodies to natalizumab in 11% of patients	1. The beneficial effect was short-lived, and persisted only through week 12 of follow-up. 2. After treatment cessation (at weeks 12-24), there was a significantly increased incidence of relapses and hospitalization in natalizumab-group compared to placebo-group
INMST Phase II	12 ms (213) [6]	+	Leukocytosis; antibodies to natalizumab in 11% of patients; increased rate of infections	1. The beneficial effect persisted only during treatment. 2. After treatment cessation (six months of follow-up), there were no significant differences between natalizumab-group and placebo-group. 3. Trial data are difficult to interpret due to striking fluctuations in the placebo-group
AFFIRM* SENTINEL	2 years (942*) (1,171) [~ 25]	+	Leukocytosis; antibodies to natalizumab in 10% of patients; increased rate of infections; two cases of PML	Data from the AFFIRM and SENTINEL trials are not published. AFFIRM is a monotherapy trial. SENTINEL is an add-on trial (natalizumab +Avenox) Both cases of PML were in SENTINEL trial

Natalizumab effects on multiple sclerosis (clinical trials)

1. Natalizumab short and long-term treatment (1–25 injections) was an unquestionably efficient restraint an active inflammatory process in multiple sclerosis (UK, INMST, AFFIRM, and SENTINEL trials).

2. All beneficial effects of natalizumab treatment persisted only during natalizumab administration on a regular basis (2 to 6 injections) or maintained for a short period of time after treatment cessation (UK and INMST trials). Stunning results, i.e., long-lasting (one year) beneficial effect after a single injection of natalizumab (Phase I trial), cannot be convincing because of the lack of consistency between data obtained in this trial and in the UK and INMST trials; the discrepancy between trial data has no satisfactory explanation. Data of follow-up periods for long-term AFFIRM and SENTINEL trials are not available.

3. After natalizumab treatment cessation, there was strong rebound effect (UK and INMST trials). Results of treatment cessation for AFFIRM and SENTINEL trials are not available.

4. Antibodies against natalizumab were detected in 10%–11% of drug-treated patients (UK, INMST, AFFIRM, and SENTINEL trials). Supposedly, anti-natalizumab antibody seropositive patients are not responsive to the treatment.

5. Common side effects associated with natalizumab treatment included increased rate of lower respiratory and urinary tract bacterial infections (UK, INMST, AFFIRM, and SENTINEL trials) and transient leukocytosis (transient after short-term treatment, UK and INMST trials; data about the long-term treatment effect on leukocytosis in AFFIRM and SENTINEL trials are not available).

6. Two cases of progressive multifocal leukoencephalopathy (PML), one of which was fatal, were reported as directly associated with two-year treatment in SENTINEL trial.

Three cases of PML in two-year trial period occurred—two cases in multiple sclerosis patients (SENTINEL trial, combination of natalizumab +Avenox) and one case in a patient with Crohn's disease (ENACT 2 trial, natalizimab alone). Scientifically speaking, there is an explanation for what happened: due to the predictable generalized immunosuppression caused by antibodies against α_4 integrins, the development of severe viral diseases (but not precisely for PML) could have been foreseen. Because multiple sclerosis is a chronic disease requiring long-term or even life-long treatment, these cases are a distress signal, calling for more adequate safety precautions during prolonged usage of integrin-antagonists.

Reduction of inflammatory process activity in multiple sclerosis after injection of only 1–6 doses of natalizumab (Tysabri) may be ground for short-term treatment at the acute stages or breakthrough forms of disease. However, given the data about the strong rebound effect after treatment cessation (UK and INMST trials), even this indication should be carefully and cautiously considered.

Reflections

Currently, integrins are emerging as objects for therapy in many human pathologies. Their widespread anatomical distribution and functional diversity make them attractive therapeutic targets for interfering in the courses of various diseases. At the same time, however any modification of integrins may create a danger of alterating a wide range of integin-regulated processes. The antibody's blockage of integrin-ligand binding causes an alteration in both bidirectional signaling mechanisms: inside-in and outside-in. The transformation of outside-in physiological signaling by anti-integrin antibodies may have two different and opposite effects on the processes that are coordinated by integrin-ligand binding. The first effect is the interruption and inhibition of outside-in mechnisms. The second effect is the possibility that the binding of integrins by antibodies may mimic the ligation of integrins by their own natural ligands and may imitate physiological signals that lead to the activation of outside-in mechanism. The transmission of false activating signals may cause an aberration in the fine-tuned control of vital cell processes—survival/apoptosis, cell cycle division, cell proliferation and differentiation, and gene expression—and may be even more dangerous than the interruption of signaling, with the most severe and unpredictable consequences.

The negative events which have been happening in the pharmaceutical industry are now very well known not only to pharmaceutical professionals, medical doctors, and scientists, but to the media, to lawyers, and to the broad population of patients. The severe adverse events associated with natalizumab (Tysabri) have deeply corroded the general trust not only in anti-α_4 integrin antibodies but also in the whole class of agents based on integrin-antagonists. The dilemma about whether they can be safely used or not may only be resolved by the accumulation and analysis of findings in different areas of science about the anatomical and functional diversity of integrins and with a real assessment of the benefit/risk ratio for integrin-antagonist drugs.

The specificity of integrin targeting should be the main issue for treatment safety.

The specificity of anti-integrin antibodies may be increased by the development of bispecific monoclonal antibody with two binding sites for different antigens: one complementary-determining region binds to the integrins, and second complementary-determining region recognizes the most individual receptor on the surface of disease-related cells. The selective delivery of the anti-integrin effectors will increase the treatment efficacy (due to the higher saturation of integrin receptors by antibodies on the surface of designated cells at the same or even lower therapeutic doses of the drug) and will reduce the side effects associated with binding antibodies to the integrins on the non-targeted cells.

Natalizumab (Tysabri) is an effective drug and, in the professional opinion of well-known neurologists, it is still considered to be a promising agent for the treatment of multiple sclerosis. To make a proper judgment about drug safety, medical doctors have to have complete information about all known aspects of the molecular mechanisms of the drug action. But until serious side effects during clinical trials with natalizumab (Tysabri) revealed its risk, the available information emphasized only one aspect of the natalizumab (Tysabri) action—its unarguable benefits for multiple sclerosis treatment. We must emphasize that the evaluation of the natalizumab (Tysabri) benefits can be done by neurologists alone. But, it is mandatory that the risk in using this agent should be assessed by a team of experts in a number of fields: immunology, infectious diseases, hematology, oncology, and embryology.

The patient-physician relationship has substantially changed during the the past decades. The patient's involvement in making treatment decisions has increased, but it is apparent that a patient cannot act as the final arbiter for his/her treatment with experimental agents. The assessment of the benefit/risk ratio still remains the sole responsibility of the doctor.

Written around 400 B.C., the ancient Hippocratic Oath delineating the physician's obligations to the medical profession and to the patients is still the keystone of ethics and values for 21st century medicine. When we, young medical doctors, take the Hippocratic Oath during the White Coat ceremony, we accept one of the most fundamental Hippocratis dictums, "Do not harm," and abide by it throughout our entire professional lives.

Vocabulary

Apoptosis—Programmed cell death. Apoptosis is the highly coordinated process that is mediated by intrinsic mechanisms, although extrinsic factors can contribute. Apoptosis is an integral part of maintaining the normal functioning of the body system. Nonsynchronized apoptosis can result in pathological outcome and is implicated in causing a variety of diseases. Excessive apoptosis can lead to the impaired growth, neurodegenerative diseases, and acquired immunodeficiency syndrome. Indiscrete suppression of apoptosis can result in autoimmune and cancer diseases. In apoptosis the earliest morphological changes occur in the nucleus of cell—condensation in the nuclear structure, nuclear shrinkage, chromatin margination and fragmentation, and breakdown of the cells into multiple spherical bodies that retain membrane integrity.

Blood brain barrier—Complex organization of the cerebral endothelial cells, pericytes and their basal lamina, which are surrounded and supported by astrocytes and perivascular macrophages. Collectively these cells separate and form the compartment that keeps most component of the blood from entering into the brain.

CD—International nomenclatures for cell-surface molecules of human lymphocytes.

- **CD4+** is expressed on the T helper cells
- **CD8+** is expressed on the cytotoxic T cells, or T killer cells

- **CD20+** is expressed on B cells
- **CD34+** is expressed on the earliest T cell precursors, or pro-T cells

Hematopoiesis—The development process by which hematopoietic stem cells give rise to all variety of specialized, mature blood cells.

Hematopoietic cell differentiation scheme—Hematopoietic cells are organised hierarchially. All blood cells derive from a common cell called the hematopoietic stem cell.

- **Hematopoietic stem cells** are multipotential cells that have the capacities to proliferate in an undifferentiated state to self-renew and differentiate into more specialized progenitor cells.

- **Hematopoietic progenitor cells** are descendent of the stem cells. Progenitor cells proliferate rapidly and differentiate into phenotypically distinct cells that give rise to the lymphoid lineage, the myeloid lineage, and the erythroid lineage, which differentiate to the lymphoid, myeloid, and erythroid precursors.

- **Hematopoietic precursor cells**—The terms "precursor" and "progenitor" cells are often used interchangeably, but sometimes the term "precursor" cells is referred to the cells with lower developmental potential than a progenitor cells. The immediate precursors of the mature blood cells have limited developmental capacity. The common lymphoid precursor divides and differentiates to give B, T, and NK cells. The common myeloid precursor divides and differentiates to give granulocytes, monocytes, dendritic cells and mast cells. The common erythroid precursor gives erythrocytes and megakaryocytes, fragments of megakaryocyte's cytoplasm called platelets.

Homeostasis—The dynamic self-regulated process that maintains relative constancy in the body internal fluids and cells. In this book, the term homeostasis is mainly referred to the process of cell homeostasis in circulatory system, which is maintained by collaborative interactions between diverse cell types and is accomplished by adhesion of the formed blood elements.

Necrosis—Mode of the cell death that is induced by severe or acute cell injury. The factors contributing to necrosis are mostly extrinsic in nature, such as osmotic, thermal, toxic, ischemic, and traumatic agents. The ultrastructural changes occur in both the cytoplasm and the nucleus: progressive loss of cytoplasmic membrane integrity, cytoplasmic swelling, disruption of the actin of the cytoskeleton, nuclear pyknosis, and finally collapse of the cell.

Opportunistic pathogens—The microorganisms causing a disease only in individuals with a reduced resistance to pathogents. In normal conditions these microorganisms are nonpathogenic.

Primary pathogens—The highly virulent microorganisms causing a disease in otherwise healthy individuals.

Abbreviations

Ab—antibody
BBA—blood brain barrier
BM—bone marrow
CAM—cell adhesion molecule
CDR—complementary determining region
CMV—cytomegalovirus
CNP—2'3'-cyclinnucleotide 2'3'-phosphodiesterase
CNS—central nervous system
CSF—cerebrospinal fluid
CTL—cytotoxic T lymphocytes
EAE—experimental autoimmune encephalomyelitis
ECM—extracellular matrix
FDA—Food and Drug Administration
EDSS—expanded disability status scale
HAMA—human anti-mouse antibody
HAART—highly active antiretroviral therapy
HSPCs—hematopoietic stem/progenitor cells
ICAM—intracellular adhesion molecule
IFN—interferon
Ig—immunoglobulin
IRD—immune restoration disease
JCV—JC virus
Mab—nomoclonal antibody
MadCam—mucosal adhesion cell adhesion molecule
MBP—myelin basic protein
MHC—major histocompatibility complex

MMP—matrix metalloproteinase
MS—multiple sclerosis
NAA—N-acetylaspartate
NK cell—natural killer cell
PB—peripheral blood
PBMC—peripheral blood mononuclear cell
PCR—polymerase chain reaction
PML—progressive multifocal leukoencephalopathy
TCR—transcriptional control region
TIMP—tissue inhibitor of metalloproteinase
VCAM—vascular adhesion molecule

References

Agostini HT, Yanagihara R, Davis V, et al. Asian genotypes of JC virus in Native Americans and in a Pacific Island population: marker of a viral evolution and human migration. *Proc Natl Acad Sci USA.* 1997;94:14542–14546

Agostini HT, Ryschkewitsch CF, Singer EJ, et al. JC virus type 2B is found more frequently in brain tissue of progressive leukoencephalopathy patients than in urine from controls *J Hum Virol.* 1998;1:200–206

Alan R, Gorska M. Lymphocytes. *J Allergy Clin Immunol.* 2003;111(2 Suppl):S476–485

Alberts B, Johnson A, Lewis J, et al. Pathogenes, infection, and innate immunity. In: Alberts B, Johnson A, Lewis J, Raff M, Roberts K, Walter P, eds. *Molecular Biology of the Cell.* 4th ed. New York, NY: Garland Science, Taylor and Francis Group; 2002: 1423–1463

Albrecht H, Hoffmann C, Degen O, et al. HAART significantly improves the prognosis of patients with HIV-associated PML. *AIDS.* 1998;12(10):1149–1154

Almeida EAC, Huovila A-PJ, Sutherland AE, et al. Mouse egg integrin functions as a sperm receptor. *Cell.* 1995;81:1095–10104

Ammons WS, Bauer RJ, Horwitz AH, et al. *In vivo* and *in vitro* pharmacology and pharmacokinetics of Human Engineering™ monoclonal antibodies to epithelial cell adhesion molecules. *Neoplasia.* 2003;5:146–154

Antinori A, Ammassari A, Giancola ML, et al. Epidemiology and prognosis of AIDS-associated PML in the HAART era. *J Neurovirol.* 2001;7:323–328

Assensi V, Carton JA, Maradona JA, et al. Progressive multifocal leukoencephalopathy associated with human immunodeficiency virus infection: the clinical, neuroimaging, virological and evolutive characteristics in 35 patients. *Med Clin (Barc)* [in Spanish]. 1999;113:210–214

Bang LM, Keating GM, Adalimumab: a review of its use in rheumatoid arthritis. *BioDrug.* 2004;18(2):121–139

Baranzini SE, Mousavi P, Rio J, et al. Transcription-based prediction of response to IFN-β using supervised computational methods. *PLoS Biol.* 2005;3(1):e2

Baum S, Ashok A, Gee G, et al. Early events in the life cycle of JC virus as potential therapeutic targets for the treatment of progressive multifocal leukoencephalopathy. *J Neurovirol.* 2003;9(Suppl1):32–37

Berengeuer J, Miralles P, Arrizabalaga J. Clinical course and prognostic factors of PML in patients treated with HAART. *Clin Infect Dis.* 2003;36(8):1047–1052

Berger JR, Levy RM, Flomenhoft D, et al. Predictive factors for prolonged survival in AIDS-associated PML. *Ann Neurol.* 1998;44(3):341–349

Berger JR. JCV-specific CD4 cell response: another piece of the puzzle in explaining some aspects of AIDS associated PML. *AIDS.* 2003,17(10):1557–1559

Bernet J, Mullick J, Singh A, et al. Viral mimicry of the complement system. *J Biosci.* 2003;28(3):249–264

Bertolottoo A, Gilli F, Sala A, et al. Evaluation of bioavailability of three types of IFN—beta in MS patients by a new quantitative-competitive-

PCR method for MxA quatification. *J Immunol Methods.* 2001;256(1–2):141–152

Bertolotto A, Gilli F, Sala A, et al. Persistent neutralizing antibodies abolish the interferon beta bioavailability in MS patients. *Neurology.* 2003;60:634–639

Blomstrand F, Aberg ND, Eriksson PS, et al. Extent of intracellular calcium wave propagation is related to gap junction permeability and level of connecxin-43 expression in astrocytes in primary cultures from four brain regions. *Neuroscience.* 1999;92:255–265

Bofill-Mas S, Pina S, Girones R. Documenting the epidemiolgic patterns of Polyomaviruses in human populations by studing their presence in urban sewage. *Appl and Environ Microbiol.* 2000;66:238–245

Boning H, Priestley GV, Nilsson LM, et al. PTX-sensitive signal in bone marrow homing of fetal and adult hematopoietic progenitor cells. *Blood.* 2004;104(8):2299–2306

Boye J, Elter T, Engert A. An overview of the current clinical use of the anti-CD-20 antibody Rituximab. *Ann Oncol.* 2003;14:520–535

Bronson RA, Fusi F. Evidence that the Arg-Gly-Asp adhesion sequence plays a role in mammalian fertilization. *Biol Reprod.* 1990;43:1019–1025

Bronson RA, Fusi F, Calzi E, et al. Evidence that a functional fertilin-like ADAM plays a role in human sperm-oolemmal interactions. *Mol Hum Reprod.* 1999;5(5):443–440

Burrows TD, King A, Loke YW. Expression of integrins by human trophoblast and differential adhesion to laminin or fibronectin. *Hum Reprod.* 1993;8:475–484

Caldarelli-Stefano R, Vago L, Omodeo-Zorini E, et al. Detection and typing of JC virus in autopsy brains and extraneural organs of AIDS patients and non-compromized individuals. *J Neurovirol.* 1999;5:125–133

Calvete JJ. Structures of integrin domains and concerted conformational changes in the bidirectional signaling mechanisms of $\alpha_{IIb}\beta_3$. *Exp Biol Med.* 2004;229:732–744

Carpenter CB, Milford EL, Sayegh MH. Transplantation in the treatment of renal failure. In: Kasper DL, Fauci AS, Longo DL, Braunwald E, Hauser SL, Jameson JL, eds. *Harrison's Principles of Internal Medicine.* 16 th ed. New York, NY: McGraw-Hill; 2005:1670–1674

Carter P, Smith L, Ryan M. Identification and validation of cell surface antigens for antibody targeting in oncology. *Endocrine-Related Cancer.* 2004;11:659–687

Caruso M, Belloni L, Sthandier O, et al. $\alpha_4\beta_1$ integrin acts as a cell receptor for murine Polyomavirus at the postattachment level. *J Virol.* 2003;77(7):3913–3921

Casoni F, Merelli E, Bedin E, et al. In serum neopterin level a marker of responsiveness to interferon-beta therapy in MS. *Acta Neurol Scand.* 2004;109(1):61–65

Chard DT, Griffin CM, Parker GJM, et al. Brain atrophy in clinical early RRMS. *Brain.* 202;125:327–337

Chatenoud L. Monoclonal antibody-based strategies in autoimmunity and transplantation. *Methods Mol Med.* 2005;109:297–328

Cheng LS, Liu AP, Yang JH. Construction, expression, and characterization of the engineered antibody against tumor surface antigen, p185$^{c-erbB-2}$. *Cell Res.* 2003;13(1):35–48

Chima SC, Ryschkewitsch CF, Stoner GL. Molecular epidemiology of human polyomavirus JC in the Biaka Pygmies and Bantu of Central Africa. *Mem Inst Oswaldo Cruz, Rio de Janeiro.* 1998;95(5):615–623

Chima SC, Agostini HT, Ryschkewitsch CF, et al. PML and JC virus genotypes in West African patients with AIDS: a pathologic and DNA sequence analysis of 4 cases. *Arch Pathol Lab Med.* 1999;123(5):395–403

Cinque P, Pierotti C, Vigano MG, et al. The good and evil of HAART in HIV-related PML. *J Neurovirol.* 2001;7:358–363

Cinque P, Koralnik IJ, Clifford DB. The evolving face of human HIV-related PML: defining a consensus terminology. *J Neurovirol.* 2003;9(Suppl 1):88–92

Clifford DB, Yiannoutsos C, Glicksman M, et al. HAART improves prognosis of HIV-associated progressive multifocal leukoencephalopathy. *Neurology.* 1999;52:623–625

Cojocaru M, Cojocaru MI, Steru O, et al. Changes in circulating immune complexes in the serum and CSF of patients with MS. *Rom J Intern Med.* 1992;30(1):51–56

Coussens LM, Werb Z. Inflammation and cancer. *Nature.* 2002;420:860–866

Dalton CM, Brex PA, Jenkins, et al. Progressive ventricular enlargement in patients with clinically isolated syndromes is associated with early development of MS. *J Neurol Neurosurg Psychiatry.* 2002;73:141–147

Dang X, Axthelm M, Letvin NL, et al. Rearrangement of Simian Virus 40 regulatory region is not required for induction of PML in immunosuppressed Rhesus Monkeys. *J Virol.* 2005;79(3):1361–1366

Daniel DC, Kinoshita Y, Khan MA, et al. Internalization of exogenous HIV-1 protein, Tat, by KG-1 oligodenroglioma cells followed by stimulation of DNA replication initiated at the JCV origin. *DNA Cell Biol.* 2004;23(12):858–867

Dasgupta S, Jana M, Lui X, et al. Role of very late antigen-4 (VLA-4) in myelin basic protein-primed T cell contact-induced expression of proinflammatory cytokines in microglia cell. *J Biol Chem.* 2003;278(25):22424–22431

Davis MJ, Wu X, Nurkiewicz TR, et al. Intergrins and mechanotransduction of the vascular myogenic response. *Am J Physiol Heart Circ Physiol.* 2001;280:H1427–H1433

Deisenhammer F, Mayringer I, Harvey J, et al. A comparative study of the relative bioavailability of different interferon beta preparations. *Neurology.* 2000;54(1):2055–2060

Del Valle L, Gordon J, Enam S, et al. Expression of human neurotropic JCV late gene product agnoprotein in human medulloblastoma. *J National Cancer Inst.* 2002;94(4):267–273

Doggrell SA. Is natalizumab a breakthrough in the treatment of multiple sclerosis? *Expert Opin Pharmacother.* 2003;4:999–1001

Dorries K. Latent and persistent polyomavris infection. In: Khalili K, GL Stoner GL. eds. *Human Polyomaviruses: Molecular and Clinical Perspectives.* New York, Wiley-Liss, Inc. 2001: 97–235

Drazen JM. Patients at risk. *New Engl J Med.* 2005;353:1–2

Du Pasquier RA, Clark KW, Smith PS, et al. JCV-specific cellular immune response correlates with a favorable clinical outcome in HIV-infected individuals with PML. *J Neurovirol.* 2001;7: 318–322

Du Pasquier RA, Corey S, Margolin DH, et al. Productive infection of cerebellar granule cell neurons by JC virus in an HIV+ individual. *Neurology.* 2003;61:775–782

Du Pasquier RA, Koralik IJ. Inflammatory reaction in PML: harmful or beneficial? *J Neurovirol.* 2003;9(Suppl 1):25–31

Du Pasquier RA, Kuroda MJ, Zheng Y, et al. A prospective study demonstrates as association between JCV-specific cytotoxic lymphocytes and the early control of PML. *Brain.* 2004;127(Pt9):1970–1978

Eash S, Tavares R, Stopa E, et. al. Differential distribution of JC virus receptor-type sialic acid in normal human tissues. *Am J Pathol.* 2004;164(2):419-427

Eisenmann DM, Kim SK. Signal transduction and cell fate specification during *Caenorhabditis elegans* vulval development. *Curr Opin Genet Dev.* 1994;4:508–516

Elphick GE, Querbes W, Jordan JA, et al. The human JCV uses serotonin receptors to infect cells. *Science.* 2004;306(5700):1380–1383

Enam S, Del Valle L, Lara C, et al. Association of human polyomavirus JCV with colon cancer. *Cancer Res.* 2002;62:7093–7101

Enam S, Sweet TM, Amini S, et al. Evidence for involvement of transforming growth factor beta 1 signaling pathway in activation of JC virus in human HIV-1-associated PML. *Arch Pathol Lab Med.* 2004;128(3):282–291

Ermekova VM, Melkonyan OS, Nanazashvili MG, et al. Immunochemical study of chromatin non-histone proteins. *Mol and Cell Biochem.* 1984a;62:133–139

Ermekova VM, Melkonyan OS, Zotova RN, et al. Chromatin regions released by endogenous nucleases are enriched in immunogenic tissue-specific proteins. *Mol Biol Rep.*1984b;9:263–267

Fazakerley JK, Walker R. Virus demyelination. *J Neurovirol.* 2003;9:148–164

Feigelson S, Grabovsky V, Winter E, et al. The Src kinase p56 up-regulates VLA-4 integrin affinity. *J Biol Chem.* 2001;276(17):13891–13901

Feldman M, Steinman L. Design of effective immunotherapy for human autoimmunity. *Nature.* 2005;435:612–619

Ferrante P, Mediati M, Caldarelli-Stefano R, et al. Increased frequency of JC virus type 2 and dual infection with JC virus type 1 and 2 in Italian PML patients. *J Neurovirol.* 2001;7:35–42

Ferrante P, Delbue S, Pagani E, et al. Analysis of JCV genotype distribution and transcriptional control region rearrangements in human HIV-positive PML patients with and without HAART. *J Neurovirol.* 2003;9(suppl 1):42–46

Freedman AS, Rhynhart K, Nojima Y, et al. Stimulation of protein tyrosine phosphorylation in human B cells after ligation of beta1 integrin VLA-4. *J Immunol.* 1993;150(5):1645–1652

Freedman MS, Blumhard LD, Brochet B, et al. and the Paris Workshop Group. International Consensus Statement on the use of disease modifying agents in MS. *Mult Scler.* 2002;8:19–23

Frenette PS, Subbarao S, Mazo IB, et al. Endothelial selectins and VCAM-1 promote hematopoietic progenitor homing to bone marrow. *Proc Nat Acad Sci USA.* 1998;95:14423–14428

Gashnault J, Kahraman M, de Goër de Herve MG, et al. Critical role of JC virus-specific CD4 T-cell responses in preventing progressive multifocal leukoencephalopathy. *AIDS.* 2003;17(10):1443–1449

Gatto B. Monoclonal antibodies in cancer therapy. *Curr Med Chem Anti-Cancer Agents.* 2004;4(5):411–414

Gazitt Y, Shaughnessy P, Liu Q. Expression of adhesive molecules on CD34+ cell in peripheral blood of Non-Hodgkin's lymphoma patients mobilized with different growth factors. *Stem Cells.* 2001;19(2):134–143

Genain CP, Fuhrman A, Menge T et al. Autoantibody reactivity in myelin/oligodendrocyte glycoproteien correlates with progressive forms of MS. *Ann Neurol.* 2002:852–866

Geschwind MD, Skolasky RI, Royal WS. The relative contribution of HAART and alpha-interferon for therapy of progressive multifocal leukoencephalopathy in AIDS. *J Neurovirol.* 2001;7:353–357

Ghosh S, Goldin E, Gordon F, et al. Natalizumab for active Crohn's disease. *N Eng J Med.* 2003;348:24–32

Giancotti FG, Ruoslahti E. Integrin signaling. *Science.* 1999;285:1028–1032

Giannelli G, De Marzo A, Scangnolari C et al. Proteolytic balance in patients with MS during interferon treatment. *J Interferon Cytokines Res.* 2002;22(6):689–692

Giffeli A, Arridge M, Jezzard P, et al. Thalamic neurodegenaration in MS. *Ann Neurol.* 2002;52:650–653

Goldberg SL, Pecora A, Alter RS, et al. Unusual virus infection (progressive multifocal leukoencephalopathy and cytomegalovirus diseases) after high-dose chemotherapy with autologous blood stem cell rescue and peritransplantation rituximab. *Blood.* 2002;99(4):1486–1488

Gonen O, Moriatry DM, Li BS. RRMS and whole brain N-acetylaspartate measurements: evidence of different clinical cohort initial observations. *Radiology.* 2002;225:261–268

Goodin DS. IFN beta therapy in MS: evidence for clinically relevant dose response. *Drugs.* 2001;61:1693–1703

Goodin DS, Frohman EM, Garmany GP. Disease modifying therapies in MS. Report of Therapeutic and Technology Assessment Subcommittee of the American Academy of Neurology and MS Council for Clinical Practice Guidelines. *Neurology.* 2002;58:169–178

Goodnow CC, Sprent J, Fazekas de St Groth B, et al. Cellular and genetic mechanisms of self tolerance and autoimmunity. *Nature.* 2005;435:590–597

Gordon FH, Clupment WY, Lai Y, et al. A randomized placebo controlled trial of humanized monoclonal antibody α_4 integrin in active Crohn's disease. *Gastroenterol.* 2001;121:268–274

Grabovsky V, Feigelson S, Chen C, et al. Second induction of alpha-4 integrin clustering by immobilized chemokines stimulates leukocytes tethering and rolling on endothelial vascular cell adhesion molecule-1 under flow condition. *J Exp. Med.* 2000;192(4):495–506

Gray F, Chrétien F, Vallat-Decouvelaere AV. The changing patterns of HIV neuropathology in the HAART era. *J Neuropathol Exp Neurol.* 2003;62(5):429–440

Greenlee JE. Progressive multifocal leukoencephalopathy—progress made and lessons relearned. *New Eng J Med.* 1998;338(19):1378–1380

Hafler DA. The distintion blurs between an autoimmune versus microbial hypothesis in multiple sclerosis. *J Clin Invest.* 1999;104:527–529

Hansson E, Rönnbäck L. Glial neuronal signaling in the central nervous system. *FASEB J.* 2003;17:341–348

Hartrich L, Weinstock-Guttman B, Hall H, et al. Dynamics of immune cell trafficking in IFN-beta treated MS patients. *J Neuroimmunol.* 2003;139(1–2):84–92

Hauser SL, Goodin DS. Multiple sclerosis and other demyelinating disease. In: Kasper DL, Fauci AS, Longo DL, Braunwald E, Hauser SL, Jameson JL, eds. *Harrison's Principles of Internal Medicine.* 16 th ed. New York, NY: McGraw-Hill; 2005:2461–2471

Hawkins SA, McDonnell GV. Benign MS? Clinical course, long-term follow-up, and assessment of prognostic factors. *J Neurol Neurosurg Psychiatry.* 1999;67:148–152

Haynes BF, Fauci AS. Introduction to the immune system. In: Kasper DL, Fauci AS, Longo DL, Braunwald E, Hauser SL, Jameson JL, eds. *Harrison's Principles of Internal Medicine.* 16 th ed. New York, NY: McGraw-Hill; 2005:1907–1930

Herdon RM. Mediacal hypothesis: why secondary progressive MS is a relentlessly progressive illness. *Arch Neurol.* 2002;59:301–304

Hoffmann C, Horst HA, Albrecht H, et al. Progressive multifocal leucoencephalopathy with unusual inflammatory response during antiretroviral treatment. *J Neurol Neurosurg Psychiatry.* 2003;74(8):1142–1144

Holman RC, Torok TJ, Belay ED, et al. Progressive multifocal leukoencephalopathy in the United States, 1979–1994: increased mortality associated with HIV infection. *Neuroepidemiol.* 1998;17(6):303–309

Hudson PJ, Souriau C. Engineered antibodies. *Nature Med.* 2003;9:129–134

Hunter T. Protein kinases and phosphotases. *Cell.* 1995:80:225–236

Hynes RO. Integrins: a family of cell surface receptors. *Cell.* 1987;48:549–554

Hynes RO. Integrins: bidirectional, allosteric signaling machines. *Cell.* 2002;110:673–687

Illes Z, Kondo T, Newcombe J, et al. Differential expresssion of NK T cell Vα24JαQ invariant TCR chain in the lesions of MS and chronic inflammatory demyelinating polyneuropathy. *J Immunol.* 2000;164:4375–4381

INFB MS Study Group. Interferon-beta is effective in RRMS. Clinical results of IFNB trial. *Neurologyl.* 1993;43:655–661

Joosten V, Lokman C, van den Hondel C, et al. The production of antibody fragments and antibody fusion proteins by yeasts and filamentous fungi. *Microb Cell Factories.* 2003, 2–1

Kalkers NF, Ameziane N, Bot JCJ, et al. Longitudinal brain volume measurement in MS: rate of brain atrophy is independent of the disease subtype. *Arch Neurol.* 2002;59:1572–1576

Kaplan RN, Riba RD, Zacharoulis S, et al. VEGFR1-positive haematopoietic bone marrow progenitors initiate the pre-metastaic niche. *Nature.* 2005;438:820–827

Kastrukoff LE, Morgan NG, Zecchini D, et al. A role for NK in immunopathogenesis of MS. *J Neuroimmunol.* 1998;86(2):123–133

Kastrukoff LE, Morgan NG, Zecchini D, et al. RRMS and NK cells. *J Neurol.* 1999;52(2):351–359

Kastrukoff LE, Lau A, Wee R, et al. Clinical relapses of multiple sclerosis are associated with "novel" valleys in NK cell functional activity. *J Neuroimmunol.* 2003;145(1–12):103–114

Katz-Brull R, Lenkinski RE, Du Pasquier RA, et al. Elevation of myoinositol is associated with disease containment in PML. *Neurology.* 2004;63(5):897–900

Kenealy SJ, Pericak-Vance MA, Haines JL. The genetic epidemiology of MS. *J Neuroimmunol.* 2003;143(1–2):7–12

Khuth ST, Akaoka H, Pagenstecher A. Morbillivirus infection of the mouse central nervous system induces region-specific upregulation on MMPs and TIMPs correlated to inflammatory cytokine expression. *J Virol.* 2001;75(17):8268–8282

Kiesseir BC, Hurtung HP. Multiple paradigm shifts in MS. *Curr Opin Neurol.* 2003;16(3):247–252

Kipriyanov SM, Le Gall F. Generation and production of engineered antibodies. *Mol Biothechnol.* 2004;26(1):39–60

Kleinschmidt-DeMasters BK, Tyler KL. Progressive multifocal leukoencephalopathy complicating treatment with natalizimab and interferon Beta-1a for multiple sclerosis. *New Engl J Med.* 2005;353:1–6

Klentzeris LD, Fishel S, McDermott M, et al. A positive correlation between expression of beta 1 integrin cell adhesion molecules and fertilization ability of human spermatozoa in vitro. *Hum Reprod.* 1995;10:728–733

Koni PA, Joshi SK, Temann UA, et al. Conditional vascular cell adhesion molecule-1 deletion in mice: impaired lymphocyte migration to bone marrow. *J Exp Med.* 2001;193(6):741–754

Kontermann RE. Recombinant bispecific antibodies for cancer therapy. *Acta Pharmacol Sin.* 2005;26(1):1–9

Koralik IJ, Du Pasquier RA, Kuroda MJ, et al. Association of prolong survival in HLA-A2+ progressive multifocal leukoencephalopathy patients with a CTL response specific for a commonly recognized JC virus epitopes. *J Immunol.* 2002;168:499–504

Kracke A, von Wussow P, Al-Masri AN. MS proteins in blood leukocytes for monitoring interferon beta-1b therapy in patients with MS. *Neurology.* 2000;54(1):193–199

Kraus J, Ling AK, Hamm S, et al. IFN-beta stabilizes blood-brain barrier characteristics of cerebral endothelial cells in vitro. *Actuelle Neurologie.* 2003:P479

Kronenberg M, Rudensky A. regulation of immunity by self-reactive T cells. *Nature.* 2005;435:598–604

Kronrwett R, Martin S, Haas R. the role of cytokines and adhesion molecules for mobilization of peripheral blood cells. *Stem Cells.* 2000;18(5):320–330

Kuhlman T, Lingfeld G, Bitsch A, et al. Acute axonal damage in MS is most extensive in early disease stages and decreases over time. *Brain.* 2002;125:2202–2212

Kurtke JF. Rating neurologic impairment in multiple sclerosis: an expanded disability scale (EDSS). *Neurology.* 1983;33(11):1444–1452

Langer-Gould A, Atkas SW, Bollen AW. Progressive multifocal leukoencephalopathy in patient treated with natalizimab. *New Engl J Med.* 2005;353:1–7

Lathrop WE, Carmichael EP, Myles DG, et al. cDNA clining reveals the molecular structure of a sperm surface protein PH-20 involved in sperm-egg adhesion and the wide distribution of its gene among mammals. *J Cell Biol.* 1990;111:2939–2949

Lewin B. Oncogenes. Gene expression and cancer. In: *Genes V*. Oxford University Press, Oxford, New York, Tokyo, 1994. pp. 1182–1201.

Li DKB, Paty DW, and UBC MS/MRI Analysis Research group and Prisms Study group. MRI results on the Prisms trial. *Ann Neurol.* 1999;46:197–206

Lu DP, Tian L, O'Neill C, et al. Regulation of cellular adhesion molecule expression in murine oocytes, peri-impantation and post-implantation embryos. *Cell Res.* 2002;12(5–6):373–383

Lublin FD, Reingold SC. Defining the clinical course of multiple sclerosis: results of an international survey. National Multiple Sclerosis society (USA) Advisory Committee on Clinical Trials of New Agents in Multiple Sclerosis. *Neurology.* 1996;46:907–911

Lucchinetti C, Brük W, Parisi J, et al. Geterogenecity of multiple sclerosis lesions: implication for the pathogenesis of demtelination. *Ann Neurol.* 2000;47(6):707–717

Magigan MT, Martinko JM, Parker J. Viruses. In: Magigan MT, Martinko JM, Parker J, eds. *Brock Biology of Microorganisms*. 8th ed. New Jersey: Prentice-Hall, Inc; 1997:248–301

Major EO, Amemiya K, Tornatore CS, et al. Pathogenesis and molecular biology of progressive multifocal leukoencephalopathy, the JC virus-induced demyelinating disease of the human brain. *Clin Microbiol Rev.* 1992;5(1):49–73

McGilvray ID, Lu Z, Bitar R et al. VLA-4 integrin cross-linking on human monocytic THP-1 cells induces tussue factor expression by mechanism involving mitogen-activated protein kinase. *J Biol Chem.* 1997;274(15):10287–10294

Miller J. Immune self-tolerance mechanisms. *Transplantation.* 2001;72(8):S5–S9

Miller DH, Khan OA, Sheremata WA, et al. A controlled trial of natalizumab for relapsing multiple sclerosis. *New Engl J Med.* 2003;348(1):15–23

Minagar A, Sheremata WA, Vollmer TL. Reduction of relapses in multiple sclerosis after anti-alpha4 integren antibody (natalizumab). *J MS Care.* 2000;3:1–6

Minagar A, Alexander JS. Blood-brain barrier disruption in MS. *Mult Scler.* 2003;9(6):540–549

Miralles P, Berenguer J, Lacruz C, t al. Inflammatory reactions in PML after HAART. *AIDS.* 2001;15(14):1900–1902

Morshed S, Mercandante M, Lombroso PJ. Genetic of childhood disorders. *J Am Acad Child Adolesc Psychiatry.* 2001:40(7):855–858

Mossakowski MJ, Zelman IB. Pathomorphological variations of the AIDS-associated PML. *Folia Neuropathol.* 2000;38(3):91–100

Mun-Bryce S, Rosenberg GA. Matrix metalloproteinases in cerebrovascular disease. *J Cerebral Blood Flow and Metab.* 1998;18(1):163–172

Munschauer FE, Hartrich LA, Srewart CC, et al. Circulating NK but not cytotoxic T lymphocytes are reduced in patients with active RRMS and little clinical disability as compared to controls. *J Neuroimmunol.* 1995;62(2):177–181

Nojima YD, Rothstein K, Sugita E, et al. Ligation on VLA-4 on T cells stimulated tyrosine phosphorylation of a 105 KDa protein. *J Exp Med.* 1992;175:1045–1053

Oger J, Freedman M. GPA Rice and the Canadian MS Clinics Network Canada. Consensus SDtatement of the Canadian MS Clinic Network on: the use of disease modifying agents in MS. *Can J Neurol Sci.* 1999;26 (4):274–275

Papadaki HA. Cell adhesion molecules in haematology. *Haema.* 1999;2(4):180–191

Papayannopoulou T, Nakamoto B. Peripheralization of hemopoitic progenitors in primates treated with anti-VLA4 integrin. *Proc Natl Acad Sci USA.* 1993;90:9374–9378

Papayannopoulou T. Mechanisms of stem/progenitor cell mobilization: the anti-VLA-4 paradigm. *Semin. Hematol.* 2000;37(Suppl.2):11–18

Papayannopoulou T. Current mechanistic scenarios in hematopoietic stem/progenitor cell mobilization. *Blood.* 2004;103(5):1580–1585

Perini P, Wadhwa M, Buttarrello M, et al. Effect of IFN-beta antibodies on NK cells in MS patients. *J Neuroimmunol.* 2000;105(1):91–95

Pfriege FW, Barres BA. Synaptic efficacy enhanced by glial cells in vitro. *Science.* 1997;277(5332):1684–1687

Pho MT, Ashot A, Atwood WJ. JC virus enters human glial cells by clathrin-dependent receptor-mediated endocytosis. *J Virol.* 2000;74(5):2288–2292

Platt JL., Lakkis FG. The road to clinical tolerance. *Transplantation.* 2001;72(8):S3–S4

Power C, Gladden JC, Halliday W, et al. AIDS—and non-AIDS-related PML association with distinct p53. *Neurology.* 2000;8(3):743–746

Pozzilli C, Romano S, Cannoni S. Epidemiology and current treatment of multiple sclerosis in Europe today. *J Rehab Res&Dev.* 2002;39(2):175–186

Prisms Study Group and the University British Columbia MS/MRI Analysis Group. Prisms-4; Long-term efficacy of IFN-beta-1a in RRMS. *Neurol.* 2001;56:1628–1636

Quesenberry PJ, Becker PS. Stem cell homing: rolling, crowling, and nesting. *Proc Natl Acad Sci USA.* 1998;95:15155–15157

Reiss K, Khalili K. Viruses and cancer: lessons from the human polyomavirus, JCV. *Oncogene.* 2003;22:6517–6523

Report of the Quality Standard Subcommittee of the America Academy of Neurology: practice advisory on selection of patients with MS for treatment with Betaseron. *Neurology.* 1994;44:1537–1540

Ricciardiello L, Laghi L, Ramamirtham P, et al. JC virus DNA sequences are frequently present in the human upper and lower gastrointestinal tract. *Gastroenterol.* 2000;119:1228–1235

Ricciardiello L, Chang DK, Laghi L, et al. Mad-1 is the exclusive JC virus strain present in human colon, and its transcriptional control region has a deleted 98-base-pair sequence in colon cancer tissues. *J Virol.* 2001;75:1996–2001

Rieckman P, Toyka KV, and the Australian-Germen-Swiss MS Therapy Consensus Group. Escalating Immunotherapy of MS. *Eur Neurol.* 1999;42:121–127

Rioux JD, Abbas AK. Paths to understanding the genetic basis of autoimmune disease. *Nature.* 2005;435:584–589

Rohwedder A, Liedigk O, Schaller J, et al. Detection of mRNA transcripts of beta 1 integrins in ejaculated human spermatozoa be nested reverse transcription PCR.. *Mol Hum Reprod.* 1996;2:499–505

Roos KL, Tyler KL. Meningitis, encephalitis, brain abscess, and empyema. In: Kasper DL, Fauci AS, Longo DL, Braunwald E, Hauser SL, Jameson JL, eds. *Harrison's Principles of Internal Medicine.* 16 th ed. New York, NY: McGraw-Hill; 2005:2484–2485

Ross JS, Gray KG, Worland PJ, et al. Anticancer antibodies. *Am J Clin Pathol.* 2003;119(4):472–485

Rothstein JD, Dykes-Hoberg M, Parbo CA, et al. Knockout of glutamate transportes reveals a major role for astroglial transport in excitotoxicity and clearance of glutamate. *Neurology.* 1996;16(3):675–686

Saarela J, Schoenberg Fejzo M, Chen D, et al. Fine mapping of a multiple sclerosis locus to 2.5 Mb on chromosome 17q22–q24. *Hum Mol Genet.* 2002;11(19):2257–2267

Safdar A, Rubocki RJ, Horvath JA. Fatal immune restoration disease in HIV type 1-infected patients with PML: impact of antiviral therapy-associated immune reconstitution. *Clin Infect Dis.* 2002;35(10):1250–1257

Saito H, Sakai H, Fujihara K, et al. PML in a patient with AIDS manifesting Gerstmann syndrome. *Tohoku J Exp Med.* 1998;186(3):169–79

Salomon DR, Mojcik CF, Chang AC, et al. Constitutive activation of integrin $\alpha_4\beta_1$ defines a unique stage of human thymocyte development. *J Exp Med.* 1994;179:1573–1584

Sanz L, Blanco B, Alvarez-Vallina L. Antibodies and gene-therapy: teaching old "magic bullets" new tricks. *Trends in Immunol.* 2004;25(4):85–91

Sariyer IK, Akan I, Del Valle L, et al. Tumor induction by simian and human polyomaviruses. *Cancer Ther.* 2004;2:85–98

Satkuman LE. Rehabilitation medicine: management of adult spasticity. *CMAJ.* 2003;25(110:1173–1179).

Sato T, Tachibana K, Nojima Y, et al. Role of VLA-4 molecule in T cell costimulation. Identification of the tyrosine phosphorylation pattern induced by the ligation of VLA-4. *J Immunol.* 1995; 195(6):2938–2947

Scarisbrick IA, Rodriquez M. Hit-hit and hit-run: viruses in playing field of MS. *Curr Neurol Neurosci Rep.* 2002;3(3):265–271

Schaller J, Glander HJ, Dethloff J. Evidence of beta 1 integrin and fibronectin interaction on spermatogenic cell in human testis. *Hum Reprod.* 1993;18:1873–1878

Schaller MD, Parson TJ. Focal adhesion kinase; an integrin-linked protein tyrosine kinase. *Trends Cell Biol.* 1993;3:258–262

Schmidbauer M, Budka H, Shah KV. Progressive multofocal leukoencephalopathy in AIDS and pre-AID era. A neuropathological comparison using immunocytochemistry and in situ hydridization for virus detection. *Acta Neuropathol.* 1990;80(4):375–380

Schumacher GA, Beebe GW, Kibler RF, et al. Problems of experimental trials of therapy in multiple sclerosis: report by the panel on the evaluation of experimental trials of therapy in multiple sclerosis. *Ann NY Acad Sci.* 1965;122:522–568

Schwartz MA. Ginsberg MH. Network and crosstalk: integrin signalling spreads. *Nature Cell Biol.* 2002;4:E65–E68

Scott LM, Priestley GV, Papayannopoulou T. Deletion of α4 integrins from adult hematopoietic cells reveals roles in homeostasis, regeneration, and homing. *Mol and Cell Biol.* 2003;23(24):9349–9360

Seth P, Diaz F, Tao-Cheng JH, et al. JC virus induces nonapoptotic death of human central nervous system progenitor cell-derived astrocytes. *J Virol.* 2004;78(9):4884–4891

Shan D, Press OW, Tsu TT, et al. Characterization of scFv-Ig constructs generated from the anti-CD-20 mAb 1F5 using linker peptides of varying lengths. *J Immunol.* 1999;162:6589–6595

Sheremata WA, Vollmer TL, Stone LA, et al. A safety and pharmacokinetic study of intravenous natalizumab in patients with MS. *Neurology.* 1999;52:1072–1074

Sheridan C. Fast track to MS drug. *Nature Biotechnol.* 2004;32(8):939–941

Sheridan C. Tysabri raises alarm bells on drug class. *Nature Biotechnol.* 2005;23(4):397–398

Shinar Y, Livnen A, Villa Y, et al. Common mutations in the familial Mediterranean fever gene associate with rapid progression to disability in non-Ashkenazi Jewish MS patients. *Genes and Immunity.* 2003;4(3):197–203

Silber E, Semra YK, Gregson NA, et al. Patients with progressive MS have elevated antibodies to neurofilament subunit. *Neurology.* 2002;58:1372–1381

Sobel RA. The extracellular matrix in multiple sclerosis: an update. *Braz J Med Biol Res.* 2001;34(5):603–609

Steinman L. Blocking adhesion molecules as therapy for multiple sclerosis: natalizuman. *Nature Rev.* 2005;4(6):810–818

Strobel BS, Möbest D, von Kleist S, et al. Adhesion and migration are differentially regulated in hematopoietic progenitor cells by cytokines and extracellular matrix. *Blood.* 1997;90(9):3524–3532

Sundström P, Svenningsson A, Nyström L, et al. Clinical characteristics of multiple sclerosis in Västerbotten County in northern Sweden. *J Neurol Neurosurg Psychiatry.* 2004;75:711–716

Swanborg RH, Whittum-Hudson JA, Hudson AP. Infectious agents and multiple sclerosis—are Chlamydia pneumoniae and human herpes involved? *J Neuroimmunol.* 2003;136(1–2):1–8

Takahashi K, Miyake S, Kondo T, et al. Natural killer type 2 bias in remission of MS. *J Clin Invest.* 2001;107:R23–R29

Tarone G, Russo MA, Hirsch E, et al. Expression of beta 1 integrin complexes on the surface of unfertilized mouse oocyte. *Development.* 1993;117(4):1369–1375

Tassie JM, Gasnault J, Bentata M, et al. Survival improvement of AIDS-related progressive multifocal leukoencephalopathy in the era of protease inhibitors. Clinical Epidemiology Group. French Hospital Database on HIV. *AIDS.* 1999;13(14):1881–1887

Tayben SB, Greer JM, Pender MP. Increased circulating antiganglioside antobodies in primary and secondary progressive MS. *Ann Neurol.* 1998;44:980–983

Theien BE, Vanderlugt CL, Eagar TN, et al. Discordant effects of anti-VLA-4 treatment before and after onset of relapsing experimental autoimmune encephalomyelitis. *J Clin Invest.* 2001;107(8):995–1006

Tienari PJ, Sumelahti ML, Rantamaki T, et al. Multiple sclerosis in western Finland : evidence for a founder effect. *Clin Neurol Neurosurg.* 2004;106(3):1750179

Tubridy N, Behan PO, Capildeo R, et al. The effect of anti-alpha-4 integrin antibody on brain lesion activity in multiple sclerosis. The UK natalizumab Study Group. *Neurology.* 1999;53(3):466–472

Ulyanova T, Scott LM, Priestley GV, et al. VCAM-1 expression in adult hematopoietic and non-hematopoietic cells is controlled by tissie inductive signals and reflects their developmental origin. *Blood.* 2005, March 15 (Epub ahead of print)

Vallittu AM, Halminen M, Peltoniemi J, et al. Neutralizing antibodies reduce MxA protein induction in interferon-beta-1a-treated MS patients. *Neurology.* 2002;58:1786–1790

Van Assche G, Rutgeerts P. Antiadhesion molecule therapy in inflammatory bowel disease. *Inflammatory Bowel Dis.* 2002;8(4):291–300

Van Assche G, van Ranst M, Sciot R, et al. Progressive multifocal leukoencephalopathy after natalizumab therapy for Chron's disease. *New Engl J Med.* 2005;353:1–7

Vermeulen M, Le Pesteur F, Gagnerault MC, et al. Role of adhesion molecules in the homing and mobilization of murine haematopoietic stem and progenitor cells. *Blood.* 1998;92(3):894–900 von Einsiedel RW, Aksamit AJ, Cornford ME, et al. PML in AIDS; a clinopathologic study and review of the literature. *J Neurol.* 1993;24(7):391–406

Vranes Z, Poljakovic Z, Marisic M. NK number and activity in MS. *J Neurol Sci.* 1989;94(1):115–123

Wagstaff AJ, Goa KL. Recombinant IFN beta 1a. A review of its therapeutic efficacy in RRMS. *Biodrugs.* 1998;10:471–494

Walsh MJ, Murray JM. Dual implication of 2'3'-cyclic nucleotide 3' phosphoesterase as major autoantigen and C3 complement-binding protein in the pathogenesis of MS. *J Clin Invest.* 1998;101:1923–1931

Wang F, Kieff E. Medical Virology. In: Kasper DL, Fauci AS, Longo DL, Braunwald E, Hauser SL, Jameson JL, eds. *Harrison's Principles of Internal Medicine.* 16 th ed. New York, NY: McGraw-Hill; 2005:1019–1027

Waubant E, Goodkin DE, Gee I, et al. Serum MMP-9 and TIMP levels are related toMRI activity in relapsing MS. *Neurology.* 1999;53(7):1397–1401

Waubant E, Gee L, Miller K, et al. IFN-beta 1a may increase serum levels of TIMP-1 in patients with relapsing remitting MS. *J Interferon Cytokine Res.* 2001;21(3):181–185

Weber T, Weber F, Petry H, et al. Immune response in PML: an overview. *J Neurovirol.* 2001;7:311–317

Weinshenker BG, Bass B, Rice GP, et al. The natural history of multiple sclerosis: a geographically based study I: clinical course and disability. *Brain* 1989a:112:133–146

Weinshenker BG, Bass B, Rice GP, et al. The natural history of multiple sclerosis: a geographically based study II: precitive value of the early clinical course. *Brain.* 1989b:112:1419–1428

Whiley DM, Mackay IM, Sloots TP. Detection and differentiation of human polyomaviruses JC and BK by LightCycler PCR.. *J Clin Microbiol.* 2001;39(12):4357–4361

Wilhams CJ, Schultz RM, Kopf GS. A role of G proteins in mouse egg activity. *Dev Biol.*1992;151:288–296

Wing MG, Moreau T, Greenwood J, et al. Mechanism of first-dose cytokine-release syndrome by CAMPATH 1-H: involvement of CD16(FcγRIII) and CD11a/CD18 (LFA-1) on NK cells. *J Clin Invest.* 1996;98(12):2819–2826

Xiong J-P, Stehie T, Goodman SL, et al. New insights into the structural basis of integrin activation. *Blood.* 2003;102:1156–1159

Yiannoutsos CT, Major EO, Curfman B, et al. Relation of JC virus DNA in the cerebrospinal fluid to survival in AIDS patients with biopsy proven PML. *Ann Neurol.* 1999;45(6):816–821

Yong VW, Chabot S, Stuve O, et al. IFN-β in the treatment of MS: mechanism of action. *Neurology.* 1998;51:682–689

Zhan XL, Hong Y, Zhu T, et al. Essential functions of protein tyrosine phosphotases Ptp2 and Ptp3 and Rim11 tyrosine phosphorylation in Saccharomyces cerevisiae meiosis and sporulation. *Mol Biol of Cell.* 2000;11(2):663–676

Zheng HY, Yasuda Y, Kato S, et al. Stability of JCV coding sequences in a case of PML in which the viral control region was rearranged markedly. *Arch Pathol Lab Med.* 2004;128(3):275–278

978-0-595-40145-1
0-595-40145-7

www.ingramcontent.com/pod-product-compliance
Lightning Source LLC
Chambersburg PA
CBHW030938180526
45163CB00002B/620